29.50

Editors

**F.W. Gehring
P.R. Halmos**

Universitext

Editors: F.W. Gehring, P.R. Halmos

Booss/Bleecker: Topology and Analysis
Charlap: Bieberbach Groups and Flat Manifolds
Chern: Complex Manifolds Without Potential Theory
Chorin/Marsden: A Mathematical Introduction to Fluid Mechanics
Cohn: A Classical Invitation to Algebraic Numbers and Class Fields
Curtis: Matrix Groups, 2nd ed.
van Dalen: Logic and Structure
Devlin: Fundamentals of Contemporary Set Theory
Edwards: A Formal Background to Mathematics I a/b
Edwards: A Formal Background to Higher Mathematics II a/b
Endler: Valuation Theory
Frauenthal: Mathematical Modeling in Epidemiology
Gardiner: A First Course in Group Theory
Godbillon: Dynamical Systems on Surfaces
Goldblatt: Orthogonality and Spacetime Geometry
Greub: Multilinear Algebra
Hermes: Introduction to Mathematical Logic
Hurwitz/Kritikos: Lectures on Number Theory
Kelly/Matthews: The Non-Euclidean, The Hyperbolic Plane
Kostrikin: Introduction to Algebra
Luecking/Rubel: Complex Analysis: A Functional Analysis Approach
Lu: Singularity Theory and an Introduction to Catastrophe Theory
Marcus: Number Fields
McCarthy: Introduction to Arithmetical Functions
Meyer: Essential Mathematics for Applied Fields
Moise: Introductory Problem Course in Analysis and Topology
Øksendal: Stochastic Differential Equations
Porter/Woods: Extensions of Hausdorff Spaces
Rees: Notes on Geometry
Reisel: Elementary Theory of Metric Spaces
Rey: Introduction to Robust and Quasi-Robust Statistical Methods
Rickart: Natural Function Algebras
Schreiber: Differential Forms
Smith: Power Series from a Computational Point of View
Smoryński: Self-Reference and Modal Logic
Stanisić: The Mathematical Theory of Turbulence
Stroock: An Introduction to the Theory of Large Deviations
Sunder: An Invitation to von Neumann Algebras
Tolle: Optimization Methods

Robert Goldblatt

Orthogonality and Spacetime Geometry

With 126 Illustrations

Springer-Verlag
New York Berlin Heidelberg
London Paris Tokyo

Robert Goldblatt
Department of Mathematics
Victoria University
Private Bag, Wellington
New Zealand

AMS Classification: 53-01, 53-B30

Library of Congress Cataloging in Publication Data
Goldblatt, Robert.
  Orthogonality and spacetime geometry.
  (Universitext)
  Bibliography: p.
  Includes index.
  1. Geometry, Affine.   2. Space and time.
3. Functions, Orthogonal.   I. Title.
QA477.G65   1987   516'.4   87-4941

Printed and bound by R.R. Donnelley & Sons, Harrisonburg, Virginia.
Printed in the United States of America.

9 8 7 6 5 4 3 2 1

ISBN 0-387-96519-X Springer-Verlag New York Berlin Heidelberg
ISBN 3-540-96519-X Springer-Verlag Berlin Heidelberg New York

To my son Tom

Preface

This book examines the geometrical notion of *orthogonality*, and shows how to use it as the primitive concept on which to base a metric structure in affine geometry. The subject has a long history, and an extensive literature, but whatever novelty there may be in the study presented here comes from its focus on geometries having lines that are self-orthogonal, or even singular (orthogonal to all lines). The most significant examples concern four-dimensional special-relativistic spacetime (Minkowskian geometry), and its various sub-geometries, and these will be prominent throughout. But the project is intended as an exercise in the foundations of geometry that does not presume a knowledge of physics, and so, in order to provide the appropriate intuitive background, an initial chapter has been included that gives a description of the different types of line (timelike, spacelike, lightlike) that occur in spacetime, and the physical meaning of the orthogonality relations that hold between them.

The coordinatisation of affine spaces makes use of constructions from *projective* geometry, including standard results about the matrix representability of certain projective transformations (involutions, polarities). I have tried to make the work sufficiently self-contained that it may be used as the basis for a course at the advanced undergraduate level, assuming only an elementary knowledge of linear and abstract algebra. Therefore I have included, where appropriate, discussion of the standard material that is being put to use. My feeling is that the topic offers an attractive way of presenting a course on geometry in a context that is rich with physical motivation, and which provides interesting content to the (already potent) ideas of projective geometry. An instructor could develop the standard material in greater depth and detail (e.g. by proving the fundamental theorem of projective geometry), as he saw fit.

One of my original motivations for this study was metamathematical: to prove that the theory of Minkowskian geometry is complete and decidable (in the sense of first-order logic). The exposition however follows the informal axiomatic style that is customary mathematical practise, and which leads to axiom sytems for spacetime and other geometries that are categorical (and would be second-order if formalised). I have deferred to an appendix an explanation of how to treat affine spaces as relational structures and obtain the metamathematical results. It would, on the other hand, be quite straightforward to include from the outset a discussion of incidence structures as models of first-order theories.

The geometry of spacetime can in fact be entirely constructed out of the single temporal relation *after* between spacetime locations, as was shown by A.A.Robb in *A Theory of Time and Space* [1914]. In a second appendix, this result is recovered, by showing that the primitive concepts used here (orthogonality, betweenness) can be defined in terms of *after*. The demonstration of this allows us also to give a simple treatment of the celebrated Alexandrov-Zeeman theorem that any light-cone preserving permutation of Minkowskian spacetime is a dilation of an inhomogeneous Lorentz transformation.

The book was written at Stanford University, during a period of sabbatical leave from the Victoria University of Wellington that was supported by both institutions and the Fulbright programme. I would like to thank Solomon Feferman and Jon Barwise for the facilities that were made available to me, and Patrick Suppes for his continuing interest in the subject. In typesetting the manuscript I was given generous access to computing equipment at Stanford's Center for the Study of Language and Information, and I am particularly grateful to the Center's Editor, Dikran Karagueuzian, whose advice and assistance in technical and aesthetic matters has considerably enhanced the final appearance of the manuscript.

Contents

1

A Trip on Einstein's Train

We begin with a story that has been told many times before, but with a version that is a little different in emphasis.

Imagine an object that moves in one dimension, such as a train moving along a railroad track. Plot spatial position along the horizontal x-axis of a two-dimensional graph, and use the vertical t-axis to plot time (Figure 1.1).

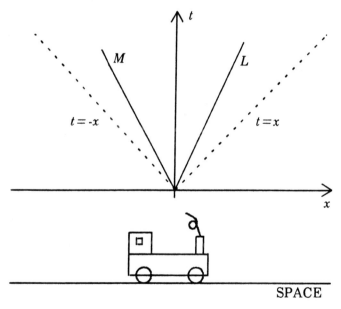

Figure 1.1

A point (x, t) on the graph represents a *spacetime location*, a particular position occupied at a certain time, and the set of all such locations occupied by a point-object (particle) is its *worldline*, or history, in spacetime.

Assume that the midpoint of the train occupies position $x = 0$ at time $t = 0$. If it remains stationary, its worldline will be the vertical axis (the line with equation $x = 0$). If it is moving to the right at a constant speed, the worldline will be a straight line of positive slope, like L in Figure 1.1. The greater is the constant speed, the more will L slope to the right, as the greater will be the increase in x relative to t. Similarly, if the train moves left at constant speed, the worldline will have negative slope, like M, sloping further to the left the greater the speed is.

Suppose that the measurement scales on the graph are chosen so that one unit along the space axis represents the distance that light travels in one unit of time. Then the broken lines of Figure 1.1 are the worldlines of photons travelling to the right ($t = x$), or left ($t = -x$), along the track from the origin $\mathbf{o} = (0, 0)$. According to relativity theory, no material object can travel as fast as light, and so the part of the train's worldline that represents its future after time $t = 0$ must lie in the region above the two broken lines, as indicated for L and M, while the region below these two lines represents possible past locations.

The two lines $t = x$ and $t = -x$ are thus referred to as *lightlike*, while lines through the origin that lie in the upper and lower regions of future and past, respectively, are *timelike*.

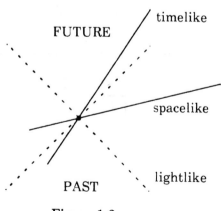

Figure 1.2

All other lines through $\mathbf{o}$ are called *spacelike*, and lie in the regions to either side. These last two regions represent locations that could belong neither to the past nor the future of the history of the midpoint of the train. It might seem natural to regard locations on the x-axis as being *simultaneous* with **o**, since they have the same time coordinate ($t = 0$) as **o**, but in fact this notion of simultaneity is a relative one. Indeed the idea of motion itself is relative: there is no absolute motion, only motion relative to some reference point. If you regard yourself as a stationary "observer" standing by the track where $x = 0$, then a train with worldline L will appear to you to be moving away. But an observer on board the train is free to regard himself as being stationary, with you and the landscape moving away from *him* (an effect seen in a film shot backwards from the rear of a moving train). The points on line L are then all associated with the same spatial position for the train-observer, but different positions for you. The *principle of relativity* states that these two viewpoints are equally acceptable. The laws of physics provide no way of distinguishing one in favor of the other.

Keep in mind that the worldlines in Figure 1.1 are not the actual paths in space of moving objects, i.e. it is not a picture that you see when you look at the physical world around you (what you see are horizontal movements along the track). The two-dimensional diagram is a pictorial *representation* of *spacetime* locations, a graphical device for attaching labels to those locations in the form of numerical coordinates. There is nothing absolute about this system of labelling. The train-observer may draw for himself a spacetime diagram from his point of view that he is stationary. The two labelling systems would then assign different labels to the same locations, and so the two pictures could not be directly superimposed on each other. Locations on the train-observer's worldline would be on a vertical line in his diagram (since he regards them all as having the same spatial coordinate), but on a line parallel to L in your diagram. Similarly, the locations in your history are represented in his diagram by a line sloping to the *left* (since that is the way he sees you as moving).

None of this is particularly surprising. What is a little harder to appreciate is that just as the assignment of spatial coordinates is relative, so too with time coordinates. All the points on the x-axis in your diagram are, for you, simultaneous - you give them all time coordinate $t = 0$. But to the observer on the train, these locations occur at different times. The locations that you judge simultaneous at any particular time all lie along a line, one that is horizontal in your diagram. The locations that are simultaneous for the moving train-observer at a given time also lie along a straight line in your diagram, but one that is at an angle to the horizontal. This can be explained by conducting another "thought-experiment" (adapted from Mermin [1968]), one that invokes the other basic hypothesis of special relativity theory, namely that the velocity of light is an absolute constant, the same for all observers regardless of what their motion is relative to the source of that light.

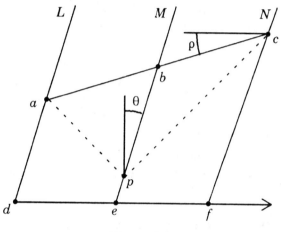

Figure 1.3

In Figure 1.3, the lines L, M, and N are the worldlines of the rear, midpoint, and front end, respectively, of a train moving at constant speed to the right. At spacetime point p, a light-source at the center of the train is switched on, and photons emitted in both directions along the train. By the principle of constancy of light-speed, the fact that the light source is moving with the train does not alter the speed that you judge these photons to have in either direction, so their worldlines in your diagram will, as before, have slopes $+1$ and -1 respectively. These photons meet the ends of the train at spacetime locations a and c, which clearly from your point of view are not simultaneous. And of course you would say that that while one one photon was travelling to the rear of the train, the rear end was travelling toward it, hence shortening the time till they met, and on the other hand the front end was moving away from the other photon, and hence delaying their collision. However the train observer regards the train as being stationary, and he knows that the light-source is positioned midway between the two ends. Thus from his point of view, the two photons appear to travel the *same* distance at the same speed to reach either end of the train, and so must arrive there at the *same* time. Therefore the two "events" *a* and *c* are judged by him to be simultaneous.

Now in Figure 1.3, it turns out that the angle $\angle bpc$ is equal to the angle $\angle bcp$, and so as $\theta + \angle bpc = \rho + \angle bcp = 45°$, it follows that angles ρ and θ have the same size, i.e. that the line *ac* makes the same angle with the horizontal that the worldlines of the train make with the vertical.

To see why $\angle bpc = \angle bcp$, observe that the segments *de* and *ef* on the space axis have the same length in the diagram, because they represent the spatial positions occupied by the ends and midpoint of the train at locations that are simultaneous for you at time $t = 0$. Since lines *ad*, *be*, and *cf* are parallel, it follows that *ab* and *bc* have the same length (because parallel lines always cut two given lines in proportional segments). But then since angle $\angle apc$ is a right angle, and the angle in a semicircle is a right angle, the circle centered on *b* and passing through *a* and *c* must also pass through *p*, showing that *bp* and *bc* have the same length, making triangle *bpc* isosceles, and giving the desired result.

This argument, with its reference to lengths and proportionality of segments, sizes of angles, angles in semicircles etc., is an exercise in the *Euclidean* geometry of the two-dimensional diagram of Figure 1.3. Reflection on the construction leads to the following conclusion.

Proposition. *Let X be an observer moving at a constant speed, with a worldline making an angle of θ to the vertical, and let p be any spacetime location. Then any other location q will be considered by X to be simultaneous with p if, and only if, the line pq makes an angle of θ to the horizontal.*

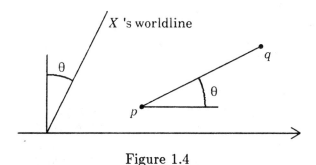

Figure 1.4

(For a more leisurely and detailed justification of this result, read Chapter 17 of Mermin [1968]).

The requirement on angles in the Proposition amounts to saying that the slope of pq must be reciprocal to that of X's worldline (i.e. if one has slope m, the other has slope $1/m$) for p and q to be simultaneous for X. Thus the locations that are simultaneous with p for X are precisely those that lie on the line through p whose slope is reciprocal to that of X's worldline. Hence if we take the same location to be the origin in both spacetime labelling systems, X's coordinatisation will be represented in your diagram as in Figure 1.5 (x', t' denote numbers assigned by X).

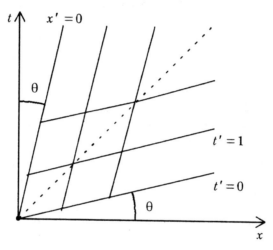

Figure 1.5

Notice that the photon with worldline equation $t = x$ in your diagram gets the same equation $t' = x'$ in X's coordinate system: the constancy of the velocity of light dictates that the two systems have the same lightlike lines.

Whenever two lines are related in such a way that one of them is the worldline of an observer moving at constant speed (from your viewpoint), and the other is a line of locations that are simultaneous for him, the two lines will be said to be orthogonal. This notion is the keystone of our study, and there is a simple algebraic formula that characterises it. Suppose L and M are a two lines that are to be determined orthogonal or not. Translate to the origin, and consider the lines L' and M' through **o** parallel to L and M, respectively, as in Figure 1.6. Then the above discussion indicates that L is orthogonal to M, denoted $L \perp M$, if, and only if, L' is orthogonal to M'. Take points $a = (x_1, t_1)$ on L' and $b = (x_2, t_2)$ on M' to serve as *direction vectors* for L and M. Put

$$a \bullet b = x_1 x_2 - t_1 t_2.$$

The number $a \bullet b$ is the *Minkowskian inner product* of a and b. In the language of matrix multiplication it can be expressed as

$$a \bullet b = (x_1 \ t_1) \begin{pmatrix} 1 & 0 \\ 0 & -1 \end{pmatrix} \begin{pmatrix} x_2 \\ t_2 \end{pmatrix}.$$

Now by the Proposition, $\mathbf{o}a$ ($= L'$) and $\mathbf{o}b$ ($= M'$) will be orthogonal just in case their slopes are reciprocal. Thus

$$\mathbf{o}a \perp \mathbf{o}b \quad \text{if and only if} \quad a \bullet b = 0,$$

so that a pair of lines are determined to be orthogonal if and only if the inner product of any two direction vectors for those lines "vanishes".

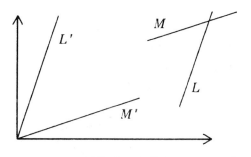

Figure 1.6

In the event that $a = b$, the inner product becomes

$$a \bullet a = x_1^2 - t_1^2,$$

and this expression can be used to characterise the three kinds of line in Figure 1.2, according to the table

$\mathbf{o}a$ is	$a \bullet a$ is
timelike	negative
lightlike	0
spacelike	positive

By these descriptions, a lightlike line is orthogonal to itself. The faster the constant speed of an observer is, the steeper will be the slope of his line of simultaneous locations, and so the closer will be the latter to his worldline (Figure 1.7). Ultimately, at the speed of light, the two lines coincide (were an observer able to travel at the speed of light, all locations in his own history would become simultaneous!).

Notice that when two *distinct* lines are orthogonal, one of them must be timelike and the other spacelike (this is not so when the number of spatial dimensions is increased, as will be seen shortly).

Timelike lines have also been called *inertia* lines, as they are the worldlines of unaccelerated particles. Spacelike lines have been called *separation* lines, since particles that occupy distinct locations on a spacelike line must be separate particles (because two such locations cannot both belong to the worldline of a single observer). Lightlike lines, being the histories of light-rays, are *optical* lines. This terminology was devised by A.A.Robb, whose book *A Theory of Time and Space* [1914] presented the first axiomatic account of spacetime of the kind to be developed here.

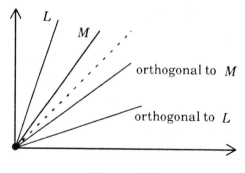

Figure 1.7

The two-dimensional geometry we have been discussing is sometimes known as the *Lorentz plane*, after the physicist H.Lorentz who developed the equations (Lorentz transformations) that relate the spacetime-labelling systems employed by two observers moving at constant speed relative to one another. Special relativity theory is concerned only with objects whose motion is of this type. The study of objects that accelerate relative to one another is the province of general relativity theory, and involves techniques from differential geometry.

Let us now lift the discussion up a dimension, and consider objects that move about in two spatial dimensions, say on some flat surface. This generates the three-dimensional spacetime of Figure 1.8. The worldlines of photons moving out from the origin in all directions on the surface (x-y plane) form a cone, as shown, known as the *lightcone*. Lines inside this cone, such as the t-axis, are timelike - the histories of observers moving at constant speeds on the plane - while lines lying outside the cone, such as the x-axis, are spacelike. Each point in this spacetime is a triple (x, y, t) of coordinates, two for spatial position, and one for time. The inner product of $a = (x_1, y_1, t_1)$ and $b = (x_2, y_2, t_2)$ is now specified to be the number

$$a \bullet b = x_1 x_2 + y_1 y_2 - t_1 t_2$$
$$= (x_1 \ y_1 \ t_1) \begin{pmatrix} 1 & 0 & 0 \\ 0 & 1 & 0 \\ 0 & 0 & -1 \end{pmatrix} \begin{pmatrix} x_2 \\ y_2 \\ t_2 \end{pmatrix},$$

and then the same description as before applies for orthogonality - $\mathbf{o}a \perp \mathbf{o}b$ iff $a \bullet b = 0$.

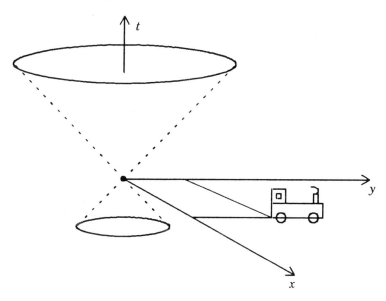

Figure 1.8

In the case $a = b$, the quantity

$$a \bullet a = x_1^2 + y_1^2 - t_1^2$$

characterises the three types of line via the same table for $a \bullet a$ as above .

An immediate divergence from the two-dimensional situation is that there are now orthogonal lines that are *both* spacelike, e.g. the x and y axes. This relates to the fact that the set of all locations that are simultaneous with a given location for a particular observer now form, not a line, but a *plane*, with all lines in this plane being orthogonal to the observer's worldline. For a stationary observer, whose worldline is vertical, this plane is horizontal (parallel to the x-y-plane). For a moving observer it is tilted, as in Figure 1.9.

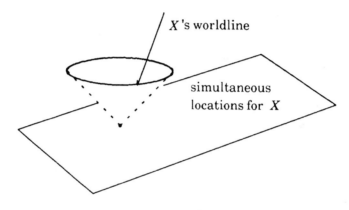

Figure 1.9

Such a plane of simultaneous locations, known as a *separation* plane, contains no lightlike lines. It consists entirely of separation lines, and its geometrical structure is like that of the familiar *Euclidean plane* (Figure 1.10).

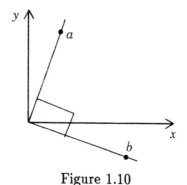

Figure 1.10

The Euclidean plane has its own inner product, usually called the *scalar* product of $a = (x_1, y_1)$ and $b = (x_2, y_2)$, given by

$$a \bullet b = x_1 x_2 + y_1 y_2$$

$$= (x_1 \ y_1) \begin{pmatrix} 1 & 0 \\ 0 & 1 \end{pmatrix} \begin{pmatrix} x_2 \\ y_2 \end{pmatrix}.$$

Euclidean lines are orthogonal ($a \bullet b = 0$) in the sense of the Euclidean inner product just in case they are *perpendicular* (at an angle of 90°) to each other.

Now in spacetime, a pair of orthogonal spacelike lines will lie in some separation plane which will itself be a plane of simultaneity for some observer. For him, the two lines will represent perpendicular directions in space, just as the x and y axes represent perpendicular directions for the stationary observer. The point is that the notion of perpendicularity of directions of motion is another observer dependent concept. To see this, imagine that an object is thrown from the train so that it moves at right angles to the track. The stationary observer sees it move in a direction perpendicular to that of the train, but the train-observer sees it fall behind him and appear to move at an obtuse angle to the direction of the train. In order for him to see the object moving at right angles to the train, it would have to be propelled along a path that the stationary observer would see as making an acute angle with the track.

Copies of the Lorentz plane also occur in three-dimensional spacetime. For example, the x-t-plane: it intersects the lightcone along a pair of optical lines (Figure 1.11).

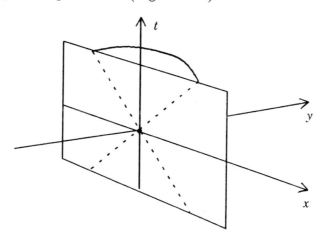

Figure 1.11

More generally, all planes containing a timelike line through the origin will have this property. They are called *inertia planes* in Robb [1914].

In addition, there is a third type of plane, called an *optical plane*, corresponding to the presence now of a third type of orthogonality relation, holding between spacelike and lightlike lines. For example, if $a = (1, 0, 0)$ and $b = (0, 1, 1)$, then $\mathbf{o}a$ is spacelike, $\mathbf{o}b$ is lightlike, and $\mathbf{o}a \perp \mathbf{o}b$. Optical planes arise as tangents to the lightcone, i.e. they intersect the lightcone along a single optical line (Figure 1.12).

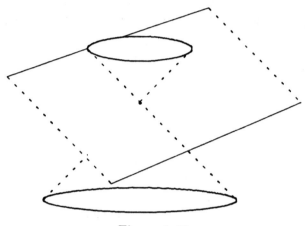

Figure 1.12

A paradigmatic two-dimensional presentation of these planes, which we shall call the *Robb plane*, is shown in Figure 1.13 as having a single self-orthogonal line through the origin.

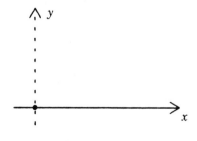

Figure 1.13

The structure of the Robb plane is encapsulated in yet another inner product, which, for $a = (x_1, y_1)$ and $b = (x_2, y_2)$ puts

$$a \bullet b = x_1 x_2$$

$$= (x_1 \ y_1) \begin{pmatrix} 1 & 0 \\ 0 & 0 \end{pmatrix} \begin{pmatrix} x_2 \\ y_2 \end{pmatrix}.$$

This time, if $a \bullet b = 0$ then one of a and b must have x-coordinate 0 and lie on the y-axis. Hence in the Robb plane the y-axis and its parallels are *singular* lines, which means that they are orthogonal to all lines that are present. The non-vertical lines are spacelike and are orthogonal to the the vertical singular lines, and only to them.

If the worldline of a photon (optical line) is orthogonal to a spacelike line, then there will be some observer who sees that photon as moving in a direction perpendicular to the spatial direction represented by the spacelike line. In the above example of $a = (1, 0, 0)$ and $b = (0, 1, 1)$, the observer is stationary, the spacelike line oa is the x-axis, and the optical line ob is the line of slope $+1$ in the t-y-plane. The photon moves in the positive direction along the y-axis.

The faster the constant speed of a moving observer is, the smaller will be the angle between his worldline and an orthogonal plane of simultaneity for him (Figure 1.9). In the limit, at the speed of light, the line lies in the plane, and is orthogonal to all lines in the plane, including itself (i.e. the plane is optical).

Having analysed in some detail the geometry of three-dimensional spacetime, it is now relatively straightforward to move by analogy to the four-dimensional context and describe the structure of the Minkowskian spacetime that represents the behaviour of objects moving about in three spatial dimensions. Here the inner product of points $a = (x_1, y_1, z_1, t_1)$ and $b = (x_2, y_2, z_2, t_2)$ becomes

$$a \bullet b = x_1 x_2 + y_1 y_2 + z_1 z_2 - t_1 t_2$$

$$= (x_1 \ y_1 \ z_1 \ t_1) \begin{pmatrix} 1 & 0 & 0 & 0 \\ 0 & 1 & 0 & 0 \\ 0 & 0 & 1 & 0 \\ 0 & 0 & 0 & -1 \end{pmatrix} \begin{pmatrix} x_2 \\ y_2 \\ z_2 \\ t_2 \end{pmatrix}$$

and the expression

$$a \bullet a = x_1^2 + y_1^2 + z_1^2 - t_1^2$$

yields the same classification of lines parallel to oa as being spacelike, timelike, or lightlike according as $a \bullet a$ is positive, negative, or zero.

The orthogonality relations holding between these lines have the same physical interpretations as in three-dimensional spacetime.

For a particular observer, the locations that are simultaneous to a given location for him now form, not a plane, but a *threefold* (three-dimensional geometry). This will be a *separation threefold*, made up entirely of spacelike lines, each of which is orthogonal to the observer's *timelike* worldline. The separation threefold will itself be a version of three-dimensional Euclidean space, as depicted in Figure 1.14.

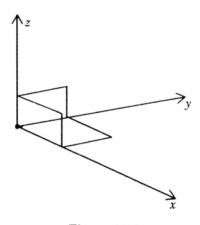

Figure 1.14

The Euclidean inner product of $a = (x_1, y_1, z_1)$ and $b = (x_2, y_2, z_2)$ is of course

$$a \bullet b = x_1 x_2 + y_1 y_2 + z_1 z_2$$

$$= (x_1 \ y_1 \ z_1) \begin{pmatrix} 1 & 0 & 0 \\ 0 & 1 & 0 \\ 0 & 0 & 1 \end{pmatrix} \begin{pmatrix} x_2 \\ y_2 \\ z_2 \end{pmatrix},$$

and "orthogonal" means "at an angle of 90°", i.e. "perpendicular" in the standard Euclidean sense.

Any threefold in four-dimensional spacetime that contains a timelike line will intersect the set of all lightlike lines (lightcone) in a three-dimensional cone of such lines, and hence will look like the three-dimensional spacetime of Figure 1.8. Indeed there will be infinitely many of these *inertia threefolds* containing any given timelike line.

There is finally a third type of threefold to consider to complete the catalogue of subgeometries of spacetime. The set of all lines that pass through a point p and are orthogonal to a given *lightlike* line L form an *optical threefold* - a three-dimensional analogue of the optical plane - in which all lines through p other than L are spacelike. Such a threefold is a version of the *Robb threefold*, depicted in Figure 1.15,

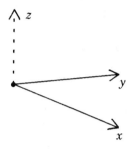

Figure 1.15

in which the z-axis is singular. The inner product characterising this geometry assigns to $a = (x_1, y_1, z_1)$ and $b = (x_2, y_2, z_2)$ the number

$$a \bullet b = x_1 x_2 + y_1 y_2$$

$$= (x_1 \ y_1 \ z_1) \begin{pmatrix} 1 & 0 & 0 \\ 0 & 1 & 0 \\ 0 & 0 & 0 \end{pmatrix} \begin{pmatrix} x_2 \\ y_2 \\ z_2 \end{pmatrix},$$

so that $a \bullet a = x_1^2 + y_1^2$, and $a \bullet a = 0$ if and only if $x_1 = y_1 = 0$.

An effective way to visualise the overall geometry of Minkowski spacetime is to consider cross-sections through the "hyperplane" with equation $t = 1$. Taking the case of three-dimensional spacetime first, intersect the lightcone with the *plane* $t = 1$. A vertical projection of this plane is shown in Figure 1.16.

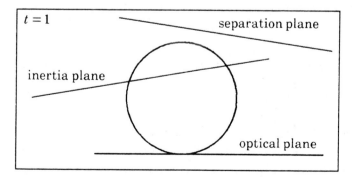

Figure 1.16

The lightcone has equation $a \bullet a = 0$, which in three dimensions is $x^2 + y^2 - t^2 = 0$. When $t = 1$, this becomes $x^2 + y^2 = 1$: the equation of a circle. Each point in the plane $t = 1$ is the cross-section of a line through the origin $\mathbf{o}$. There are three types of lines in the plane $t = 1$, arising as cross-sections of the three types of plane in spacetime. Lines exterior to the circle represent separation planes, lines tangent to the circle represent optical planes having a single lightlike line through $\mathbf{o}$, and lines meeting the circle in two points represent inertia planes (copies of the Lorentz plane).

Moving up to four dimensions, where the lightcone has equation $x^2 + y^2 + z^2 - t^2 = 0$, the equation $t = 1$ now defines a threefold which intersects the lightcone where $x^2 + y^2 + z^2 = 1$: the equation of a sphere (Figure 1.17).

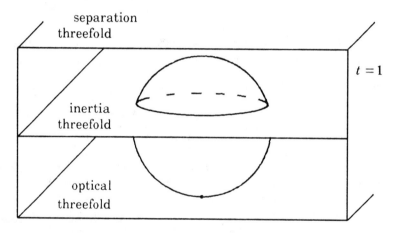

Figure 1.17

The points, lines, and planes of this threefold are the cross-sections of the lines, planes, and threefolds that pass through the origin in Minkowski spacetime (note that the center of the sphere is $(0,0,0,1)$, not the origin $(0,0,0,0)$). Points interior to the sphere represent timelike lines, points on the sphere represent lightlike lines, and exterior points represent spacelike lines. As before there are three types of line, meeting the sphere in 0, 1, or 2 points, and being the cross-sections of separation, optical, and inertia planes, respectively. In addition there are three types of plane in the threefold $t = 1$:

- planes external to the sphere, being cross-sections of separation threefolds (Figure 1.14);

- planes tangential to the sphere, being cross-sections of optical threefolds (Figure 1.15), with the unique point of contact with the sphere representing the unique optical line through o in the optical threefold; and

- planes cutting the sphere in a circle, being cross-sections of inertia threefolds (Figure 1.8), with the circle representing the lightcone of the threefold.

In this chapter it has been shown that the word "orthogonal" has a variety of meanings which are themselves characterised algebraically by a variety of interpretations of "inner product". Our aim in what follows will be to develop a coordinate-free axiomatic account of Minkowskian geometry. Beginning with an abstract set of points and lines, and a binary relation between these lines, the circumstances under which this relation may be regarded as one of *orthogonality* will be revealed by eliciting axioms that capture the essence of the notion, i.e. the properties that are common to all its interpretations (e.g. in each case above we had symmetry: if $L \perp M$ then $M \perp L$). Then by considering properties of the orthogonality relation (presence of self-orthogonal or singular lines etc.) that distinguish its interpretations, categorical axiomatic descriptions will be obtained for all the geometrical structures that have been discussed (and some others besides). In other words, in each case a set of axioms will be found that has that particular geometry as its one and only model.

2

Planes

In order to carry out the program outlined at the end of Chapter 1, we will retrace the path of that chapter, giving first a systematic analysis of planes, which is then used to discuss threefolds, and finally Minkowski spacetime itself. The present chapter is devoted to a general study of the two-dimensional situation.

2.1 Affine Planes and Fields

The Euclidean, Lorentz, and Robb planes described in Chapter 1 are distinguishable by their orthogonality relations and their inner products, but as far as points and lines are concerned they look identical. This common structure is known as the *real affine plane*, denoted $\alpha_{\mathbf{R}}$. Its points are the pairs (x, y) of real numbers, while its lines are the solution sets of linear equations, i.e. equations of the form

$$lx + my + n = 0,$$

with at least one of l, m being non-zero. A point (x_0, y_0) lies on the line determined by this equation if, and only if, it is a solution of the equation, i.e. $lx_0 + my_0 + n = 0$.

The abstract notion of an affine plane is presented as follows. An *incidence structure* is a triple $\alpha = (\mathcal{P}, \mathcal{L}, \mathcal{I})$, where

- $\mathcal{P}$ is a set, whose members are called *points*;
- $\mathcal{L}$ is another set, disjoint from $\mathcal{P}$, whose members are called *lines*; and
- $\mathcal{I}$ is a binary relation from $\mathcal{P}$ to $\mathcal{L}$, i.e. $\mathcal{I} \subseteq \mathcal{P} \times \mathcal{L}$.

If $a\mathcal{I}L$, where a is a point and L is a line, it will be said that a is *incident to* L, that a *lies on* L, or that a *passes through* L, etc. Two lines L, M will be called *parallel*, denoted $L \parallel M$, if either $L = M$, or else there is no point that lies on both of them. A set of points is *collinear* if there is a single line passing through all of them, while a set of lines is *concurrent* if there is a single point through which they all pass.

An incidence structure is an *affine plane* if it satisfies the following three axioms.

A1. *Any two distinct points lies on exactly one line.*

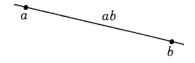

Figure 2.1

(the unique line passing through a and b is denoted ab).

A2. *There exist at least three non-collinear points.*

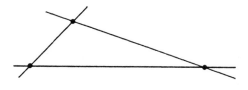

Figure 2.2

A3. *Given a point a and a line L, there is exactly one line M that passes through a and is parallel to L.*

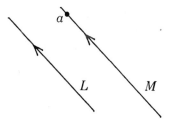

Figure 2.3

The relation $\parallel$ of parallelism is an equivalence relation (reflexive, transitive, and symmetric) on $\mathcal{L}$, and its equivalence classes (i.e. the sets $\{M : L \parallel M\}$ for each line L) are called *parallelism classes*. Two lines L, M that are distinct and not parallel must meet at exactly one point, which is denoted $L \cap M$, for by A1, lines with two distinct points in common cannot be distinct lines.

The lines of an incidence structure are often taken to be subsets of $\mathcal{P}$. This can always be effected by replacing L by the set $L' = \{a : a\mathcal{I}L\}$ of all points that lie on L, in which case incidence $(a\mathcal{I}L)$ becomes set membership $(a \in L')$. It will however be convenient at times to let $\mathcal{I}$ be a more general relation.

Desarguesian Planes
An affine plane is *Desarguesian* if it satisfies the

AFFINE DESARGUES PROPERTY: *if a, b, c and $a', b'c'$ are two triangles (i.e. non-collinear triples of points) having the lines aa', bb', and cc' concurrent, and if $ab \parallel a'b'$ and $ac \parallel a'c'$, then $bc \parallel b'c'$.*

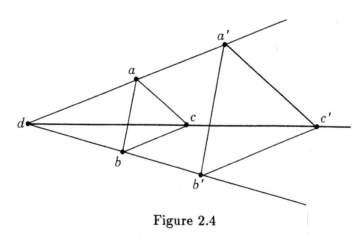

Figure 2.4

A consequence of the Desargues property is the following proposition, which will prove useful latter on.

LITTLE AFFINE DESARGUES PROPERTY: *if a, b, c and a', b', c' are triangles such that $aa' \parallel bb' \parallel cc'$, and $ab \parallel a'b'$ and $ac \parallel a'c'$, then $bc \parallel b'c'$ (Figure 2.5).*

In the wider context of projective geometry (cf. Chapter 3), Figure 2.5 maybe seen as the special case of Figure 2.4 in which d is taken as a point "at infinity". It is however a straightforward exercise to derive the Little Desargues property in any Desarguesian affine plane (Behnke et.al. [1974], p.71).

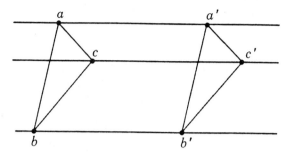

Figure 2.5

Pappian Planes

A pair L, M of lines in an affine plane has the PAPPUS PROPERTY if whenever a, b, c is a triple of points on L, and a', b', c' is a triple on M such that $ab' \parallel a'b$ and $ac' \parallel a'c$, then $bc' \parallel b'c$.

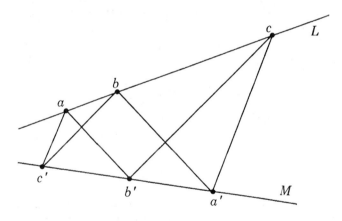

Figure 2.6

If any of a, b, c lies on M, then the hypothesis $ab' \parallel a'b$ and $ac' \parallel a'c$ implies that the other two lie on M as well (using A3), so that $L = M$ and $bc' \parallel b'c$ follows. Thus the Pappus property is only at issue in the event that no point of one triple is collinear with the other triple.

An affine plane is *Pappian* if every pair of its lines has the Pappus property. *Such a plane is always Desarguesian.* The real affine plane is Pappian, but there are many other examples besides. In order to establish the Pappus property in α_R it suffices to know that the real-number system **R** has the structure of a *field*.

Field-planes

A *division ring* is a set D with two binary operations + (addition) and · (multiplication), and elements 0 (zero) and 1 (unity), such that
- D is an Abelian group under +, with identity 0;
- $D - \{0\}$ is a group under ·, with identity 1;
- The distributive laws

$$x \cdot (y + z) = x \cdot y + x \cdot z, \quad (y + z) \cdot x = y \cdot x + z \cdot x$$

  hold for all $x, y, z \in D$.

The additive inverse of x in D is denoted by $-x$, while the multiplicative inverse of $x \neq 0$ is denoted x^{-1} or $1/x$. The element $x \cdot y^{-1}$ is often written x/y, or $\frac{x}{y}$.

A division ring in which multiplication is commutative ($x \cdot y = y \cdot x$) is called a *field*. The set **R** of real numbers is a field under the usual arithmetical operations of addition and multiplication.

Given a division ring D, the affine plane α_D over D is defined by exact analogy with $\alpha_{\mathbf{R}}$. The points of α_D are the pairs (x, y) of elements of D. The lines are the triples (l, m, n) with either $l \neq 0$ or $m \neq 0$. Point (x, y) is incident with line (l, m, n) just in case

$$l \cdot x + m \cdot y + n = 0.$$

This construction makes α_D into a *Desarguesian* affine plane, which is Pappian if, and only if, multiplication in D is commutative, i.e. iff D is a field.

Isomorphism

When may two incidence structures $(\mathcal{P}, \mathcal{L}, \mathcal{I})$ and $(\mathcal{P}', \mathcal{L}', \mathcal{I}')$ be regarded as being essentially the same geometry? Answer: when there exists an *isomorphism* between them. An isomorphism is a one-to-one mapping from $\mathcal{P}$ to $\mathcal{P}'$ that preserves collinearity, which means that whenever S is a collinear set of points in $\mathcal{P}$, then $f(S) = \{f(a) : a \in S\}$ is a collinear subset of $\mathcal{P}'$, and conversely. What this amounts to is that for each line L in $\mathcal{L}$, $f(L) = \{f(a) : a\mathcal{I}L\}$ is (the set of all points that lie on) a line in $\mathcal{L}'$. An isomorphism may alternatively be characterised as consisting of a one-to one function f from $\mathcal{P}$ onto $\mathcal{P}'$, and a one-to-one function F from $\mathcal{L}$ onto $\mathcal{L}'$, such that in general a lies on L if, and only if, $f(a)$ lies on $F(L)$ (Stevenson [1972], §1.5).

An isomorphism provides an exact matching of points and lines in one structure to points and lines in the other in such a way that corresponding elements exhibit identical incidence properties. The two geometries are indistinguishable by any geometrical properties to do with incidence, and so may be regarded as manifestations of one and the same abstract form. Thus if two structures are isomorphic, then one will be Desarguesian or Pappian, say, precisely if the other is. And if all structures that have a certain property happen to be isomorphic, then it will be said that there is "essentially only one" structure with that property.

To illustrate the isomorphism concept, take any incidence structure $(\mathcal{P}, \mathcal{L}, \mathcal{I})$, replace each line L by the set $L' = \{a \in \mathcal{P} : a\mathcal{I}L\}$, and put $\mathcal{L}' = \{L' : L \in \mathcal{L}\}$. Then the identity function on $\mathcal{P}$ is an isomorphism from the original structure onto $(\mathcal{P}, \mathcal{L}', \in)$. This substantiates the earlier claim that we may always regard a line as being a set of points, and incidence as being set-membership.

The Coordinatisation Problem

The coordinatisation problem for a given plane α is the question as to whether the points of α can be represented as pairs (x, y) of coordinates taken from some algebraic structure. Each type of algebraic structure produces its own coordinatisation problem. For fields the problem is: which affine planes are isomorphic to ones of the form α_F, where F is a field? The answer to that is: precisely the *Pappian* ones. Each Pappian affine plane can be coordinatised by a field.

Conversely, each type of geometrical structure gives rise to a problem about the kind of algebraic structure that is needed to represent it. For instance, the coordinatisation problem for Desarguesian planes is solved by the notion of a division ring. A plane is Desarguesian if, and only if, if is isomorphic to the plane over some division ring.

Now let α be a Pappian affine plane. To coordinatise α, proceed as follows. Let L and M be two lines meeting at a point o. Take a point e on L distinct from o, and a point $e' \neq o$ on M (Figure 2.7: such a configuration must exist by A3).

Let F be the set of all points that lie on L. To define addition in F, use the line N through e' that is parallel to L. Given $a, b \in F$, let the line through b parallel to M meet N at b'', and then let the line through b'' parallel to $e'a$ meet L at c. Declare $a + b = c$ (Figure 2.8).

The intuitive basis of this construction can be seen by imagining that L an M are the x and y axes in the Euclidean plane. From

the parallelograms $obb''e'$ and $acb''e'$ it follows that the line segment
from o to b is of the same length, denoted $|ob|$, as that from e' to b'',
and that $|e'b''| = |ac|$, so that $|ob| = |ac|$. But then

$$|oc| = |oa| + |ac| = |oa| + |ob|.$$

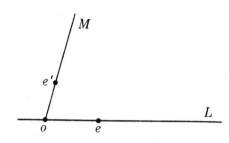

Figure 2.7

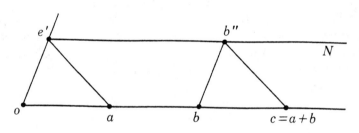

Figure 2.8

To define $a \cdot b$, let the line through b parallel to ee' meet M at b',
and then let the line through b' parallel to $e'a$ meet L at c. Declare
$a \cdot b = c$ (Figure 2.9).

Illustrating with the Euclidean plane as above, suppose that
$|oe| = |oe'| = 1$. The given construction makes the triangles oee'
and obb' similar, with corresponding sides being proportional. Thus

$$\frac{|ob|}{|oe|} = \frac{|ob'|}{|oe'|},$$

and so $|ob| = |ob'|$. But also oae' and ocb' are proportional, so

$$\frac{|oc|}{|oa|} = \frac{|ob'|}{|oe'|} = |ob|,$$

and hence $|oc| = |oa| \cdot |ob|$.

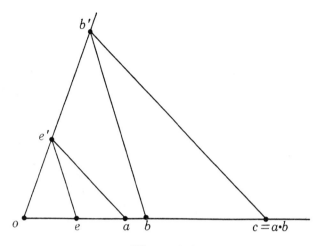

Figure 2.9

Using only the fact that α is Desarguesian, it can be shown that these definitions of $+$ and $\cdot$ make F into a division ring, with the point o as the additive identity, and e as the multiplicative identity. Moreover, the construction does not depend on the particular lines and planes used, in the sense that if the definitions are repeated with some other non-collinear triple in place of o, e, e', then the result will be a division ring isomorphic to F. In order then to coordinatise α, observe that the assignment of b' to b as in Figure 2.9 establishes a bijective correspondence between F and the set of points on M (which takes e to e' and leaves o fixed). Then given a point a in the plane, let the lines through a parallel to M and L meet L at x and M at y', respectively (Figure 2.10).

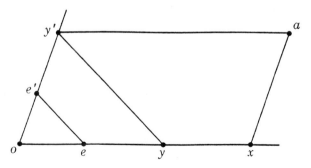

Figure 2.10

Let y correspond to y' in the bijection just referred to (i.e. $yy' \parallel ee'$). Then a is assigned the coordinate pair (x, y). (The motivation for this should be apparent.)

The function $a \mapsto (x, y)$ is an isomorphism from α onto the plane α_F. Notice that $x \in F$ receives the coordinates $(x, 0)$, while the point y' on M becomes $(0, y)$. Thus L becomes the "x-axis" and M the "y-axis" in α_F.

The procedure just outlined establishes that the Desarguesian planes are coextensive with the planes over division rings. To complete the analysis in relation to Pappian planes, the Pappus property is invoked at the final stage to show that multiplication in the coordinate set F is commutative. This is indicated in Figure 2.11.

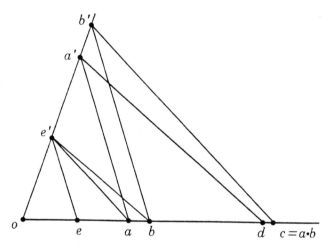

Figure 2.11

Here c is the point $a \cdot b$ as constructed in Figure 2.9, and d is the point where L meets the line through a' parallel to $e'b$, so that $d = b \cdot a$ by definition. Consider the triples e', a', b' on M and d, b, a on L. The construction gives $e'b \parallel a'd$ and $a'a \parallel b'b$, so that if the pair L, M has the Pappus property it must follow that $e'a \parallel b'd$. But $e'a \parallel b'c$, so $b'd$ and $b'c$ are parallel, and hence, having b' in common, are the same line. Thus d lies on $b'c$ and on L, i.e. $d = c$, giving $b \cdot a = a \cdot b$ as desired.

There are two points to note about this coordinatisation procedure that will be important later.

1. The choice of the lines L and M at the outset was quite arbitrary. Thus in coordinatising a Desarguesian plane, any pair

of intersecting lines may be selected to serve as the coordinate axes.

2. To establish that multiplication was commutative, the Pappus property was needed only for the pair of coordinate axes. But that guaranteed that α was isomorphic to the field-plane α_F, and hence was fully Pappian. In view of point 1, it follows that in order for a Desarguesian plane to be Pappian, it suffices that it have at least one pair of intersecting lines with the Pappus property.

Equations and Slopes of Lines

Whenever convenient, a Pappian affine plane may be taken to be presented in the form α_F for some field F. The properties of a field ensure that the basic analytic geometry of α_F works in very close analogy to that of the real affine plane $\alpha_\mathbf{R}$. Each line is given by an equation $lx + my + n = 0$, where $(l, m) \neq (0,0)$. If $m = 0$, then l^{-1} exists and can be used to transform this into an equation of the form

$$x = c,$$

for some $c \in F$. Lines with equations of this type will be called *vertical*, since they are precisely those lines that are parallel to the y-axis (the latter being the line $x = 0$). If $m \neq 0$, then the equation can be put in the form

$$y = s \cdot x + c. \tag{†}$$

The "number" s is then called the *slope* of the line in question. Two lines in α_F prove to be parallel if, and only if, they have the same slope. Thus the parallelism class of the line (†) is the set of all lines of the form $y = s \cdot x + d$, for all $d \in F$. (†) passes through the point $(0, c)$, and hence the line through $(0,0)$ parallel to (†) has the equation $y = s \cdot x$ and contains the point $(1, s)$ (Figure 2.12). In particular, the parallelism class of the x-axis is the set of "horizontal" lines with slope 0, i.e. the lines $y = c$ for all $c \in F$.

This description gives a complete classification of the parallelism classes of α. They comprise the set of all vertical lines as one class, together with, for each $s \in F$, the class of all lines of slope s. Each parallelism class has as a canonical representative its one and only member that passes through the origin $(0,0)$. This canonical line has equation $y = s \cdot x$, or $x = 0$ in the case of the class of vertical lines.

If the line joining the points (x_1, y_1) and (x_2, y_2) is not vertical, then $x_1 \neq x_2$, and the slope of this line is

$$\frac{y_2 - y_1}{x_2 - x_1}.$$

The notion of the slope of a line will play a central role in the coordinatisation of orthogonality relations in §2.6.

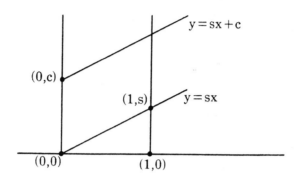

Figure 2.12

Collineations and Coordinate Changes
An isomorphism from a plane onto itself is known as a *collineation*. One way to obtain a collineation when α is the field-plane α_F is to select elements $a, b, c, d, e, f \in F$ with $a \cdot e \neq b \cdot d$ and put

$$x' = a \cdot x + b \cdot y + c$$
$$y' = d \cdot x + e \cdot y + f.$$

The collineation then takes (x, y) to (x', y') and thereby effects a *coordinate change*. There are two ways of viewing this procedure:

1. The coordinate system for α is seen as remaining fixed, with the collineation moving points to different places. The point with coordinates (x, y) is moved to the point (x', y') in the same coordinate system. This is called an *alibi* transformation.

2. The points of α remain fixed where they are, while the coordinate system is changed. No points are moved, so the change of coordinates is a change of the label that *names* a particular point. The point with label (x, y) receives the new name (x', y'). This is called an *alias* coordinate change.

Fano Planes
The smallest of all fields is $\mathbf{Z}_2 = \{0, 1\}$, with the usual arithmetical

meaning (modulo 2) of $+$ and $\cdot$ for the numbers 0 and 1. The plane over $\mathbf{Z}_2$ is depicted in Figure 2.13.

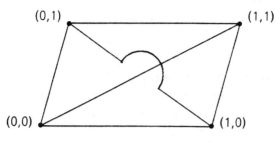

Figure 2.13

There are four points, with three lines through each point. The kink in the diagonal is there to emphasis that the two diagonals do not have a point in common. Thus this little plane has a parallelogram whose diagonals are parallel! (which shows how misleading it can be to try to visualise the structure of some planes on real affine paper).

Now the four points in Figure 2.13 are present in any field-plane, and so we ask: for which planes do they comprise a parallelogram with parallel diagonals? In fact it can be shown that for any α_F the following are equivalent.

1. Every parallelogram has parallel diagonals.

2. There exists a parallelogram with parallel diagonals.

A plane satisfying this property is called a *Fano* plane.

Notice that the points (1,0) and (0,1) of Figure 2.13 are the points e and e' of Figure 2.7, while (1,1) is the point e'' of Figure 2.8 when $b = e$. Thus Figure 2.13 is the configuration used in Figure 2.8 to define $e + e$, i.e. $1 + 1$. In fact $e + e$ is the point where the line through e'' parallel to ee' meets the x-axis. But if the diagonals are parallel, then this line is just the diagonal oe'', so that $e + e$ is the point o, i.e. $1 + 1 = 0$. And conversely.

A field (like $\mathbf{Z}_2$) in which $1 + 1 = 0$ is said to be of *characteristic* 2 (such a field actually has $x + x = 0$, hence $x = -x$, for all x). Thus α_F is a Fano plane if, and only if, F has characteristic 2. By a quirk of nomenclature, the name *Fano's axiom* is customarily given to the statement

the diagonals of any parallelogram intersect,

and so a Fano plane is one that does not satisfy Fano's axiom.

Since for each positive integer n there exists a field of characteristic 2 with 2^n members, it follows that there are infinitely many (finite) Fano planes. Most algebraic discussions of orthogonality exclude fields of characteristic 2 from the outset. We will find it possible to include them fully in our axiomatic classifications, although their behaviour is relatively pathological, as will be seen.

Full details of the theory sketched in this section may be found in numerous textbooks, such as the ones cited in the Bibliography by Behnke et. al., Blumenthal, Ewald, Garner, Hartshorne, Mihalek, Seidenberg, Stevenson, and Szmielew.

2.2 Metric Vector Spaces

The points $a_i = (x_i, y_i)$ of a field-plane α_F are also the vectors of the canonical two-dimensional vector space over F, which has the basic operations

$$a_i + a_j = (x_i + x_j, y_i + y_j)$$
$$\lambda a_i = (\lambda \cdot x_i, \lambda \cdot y_j)$$

of *vector addition*, and *scalar multiplication*, respectively.

Abstractly, a *vector space* over the field F is an Abelian group $(V, +, \mathbf{o})$ with an operation $(\lambda, a) \mapsto \lambda a \in V$ of multiplication of vector $a \in V$ by scalar $\lambda \in F$ satisfying the laws

$$\lambda(a + b) = \lambda a + \lambda b,$$
$$(\lambda + \mu)a = \lambda a + \mu a,$$
$$(\lambda \cdot \mu)a = \lambda(\mu a),$$
$$1a = a$$

(the reader is assumed to be familiar with the elementary theory of vector spaces).

Any vector space gives rise to an incidence structure in which the points are the vectors (members of V), and lines are defined as follows. If a and b are vectors with b non-zero ($b \neq \mathbf{o}$), then the set of vectors

$$L = \{a + \lambda b : \lambda \in F\}$$

is defined to be the *line through a in the direction of b* (Figure 2.14: putting $\lambda = 0$ shows $a \in L$).

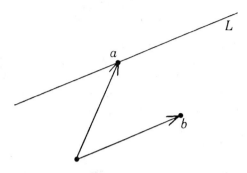

Figure 2.14

This line may be presented in the *parametric* form $L : a + \lambda b$, in which the expression $a + \lambda b$ takes all points of L as values as the parameter λ ranges over all members of F. Any scalar multiple λb of b is a *direction vector* for L (so that a scalar multiple of a direction vector for L is itself a direction vector for L).

Relating this back to the field-plane α_F, a typical point on the line with equation $y = s \cdot x + c$ has the form

$$(\lambda, s \cdot \lambda + c) = (0, c) + \lambda(1, s),$$

while a typical point on the line $x = c$ is

$$(c, \lambda) = (c, 0) + \lambda(0, 1).$$

If c and d are points on a line L, then the vector $(d - c)$ is a direction vector for L:

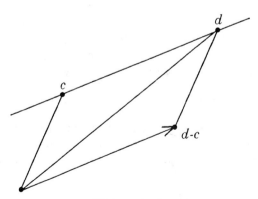

Figure 2.15

(as is $(c - d)$ of course), and so the unique line through c and d has the expression $c + \lambda(d - c)$ as a parametric form.

Now in Chapter 1, the name *inner product* was given to a function assigning a number (scalar) to a pair of points (vectors), and several examples of this idea were given in spaces of different dimensions. Abstractly, an inner product on a vector space V over a field F is defined to be a function that assigns a scalar $a \bullet b \in F$ to each pair (a, b) of vectors, and is

1) *bilinear*, i.e. the laws

$$a \bullet (\lambda b + \mu c) = \lambda \cdot (a \bullet b) + \mu \cdot (a \bullet c)$$

$$(\lambda b + \mu c) \bullet a = \lambda \cdot (b \bullet a) + \mu \cdot (c \bullet a)$$

are satisfied, and

2) *symmetric*, i.e.

$$a \bullet b = b \bullet a.$$

(Note: the symmetry condition implies that each bilinearity equation is deducible from the other.)

A vector space provided with an inner product is called a *metric vector space*. In such a structure, vectors a and b are called *orthogonal* if $a \bullet b = 0$. A non-zero vector a is *null*, or *isotropic*, if it is self-orthogonal, i.e. if $a \bullet a = 0$, and is *singular* if it is orthogonal to all vectors, i.e. if $a \bullet b = 0$ for all $b \in V$. A metric vector space is a *nonsingular space* if it has no singular vectors, and an *anisotropic space* if it has no null vectors. It is a *null space* if all of its vectors are orthogonal to each other, which is equivalent to saying that all of its vectors are singular.

Let $L : a + \lambda l$ and $M : c + \mu m$ be two lines in a metric vector space, with direction vectors l and m respectively. Then L is *orthogonal* to M, denoted $L \perp M$, if vector l is orthogonal to vector m, i.e. $l \bullet m = 0$. This definition is independent of the choice of direction vectors, since if $l \bullet m = 0$, then for any λ and μ,

$$(\lambda l) \bullet (\mu m) = \lambda \cdot \mu \cdot (l \bullet m) = \lambda \cdot \mu \cdot 0 = 0,$$

so that λl and μm are orthogonal vectors. Thus it is appropriate to define L to be a *null* line if $L \perp L$, and a *singular* line if $L \perp M$ holds for all lines M in the vector space V. A *nonsingular* line is one that is not singular.

Theorem 2.2.1. *$L \perp M$ implies $M \perp L$.*

Proof. $l \bullet m = 0$ implies $m \bullet l = 0$.

Theorem 2.2.2. *If a, b, c, and d are distinct points in V such that line ab is orthogonal to line cd, and ac is orthogonal to bd, then ad is orthogonal to bc.*

Proof. Since any line uv has $(u - v)$ and $(v - u)$ as direction vectors, $ab \perp cd$ yields

$$(b - a) \bullet (d - c) = 0,$$

which expands, using bilinearity, to

$$b \bullet d - a \bullet d - b \bullet c + a \bullet c = 0. \tag{1}$$

Similarly, $ac \perp bd$ implies

$$c \bullet b - a \bullet b - c \bullet d + a \bullet d = 0. \tag{2}$$

Adding (1) and (2) and applying the symmetry of the inner product leads to

$$d \bullet b - a \bullet b - d \bullet c + a \bullet c = 0,$$

and so

$$(d - a) \bullet (b - c) = 0,$$

i.e. $ad \perp bc$ as desired. $\qquad\qquad\qquad\qquad\qquad\qquad\quad\square$

The whole point of taking an abstract approach here is that a single proof will show once and for all that results like the ones just obtained hold in all metric vector spaces, regardless of their dimension and independently of any particular formula for calculating inner products. The next result (Lemma 2.2.3) is also general, but requires a more concrete presentation in terms of coordinates.

Recall that the canonical n-dimensional vector space over F is the set F^n of all n-tuples $(x_1, \ldots, x_n)$ of scalars $x_i \in F$, with the operations

$$(x_1, \ldots, x_n) + (y_1, \ldots, y_n) = (x_1 + y_1, \ldots, x_n + y_n)$$
$$\lambda(x_1, \ldots, x_n) = (\lambda \cdot x_1, \ldots, \lambda \cdot x_n).$$

If V is any vector space over F of dimension n, then V is isomorphic to F^n. V has an n-element *basis*, i.e. a set of vectors $\{v_1, \ldots, v_n\}$ such that for any $a \in V$ there is a unique n-tuple $\mathbf{a} = (\lambda_1, \ldots, \lambda_n)$ of scalars satisfying

$$a = \lambda_1 v_1 + \cdots + \lambda_n v_n.$$

The assignment $a \mapsto \mathbf{a}$ then gives a vector space isomorphism between V and F^n, and provides a "coordinatisation" of the vectors in V.

Suppose further that V is a metric vector space, and consider the inner product $a \bullet b$ of two vectors with coordinates $\mathbf{a} = (\lambda_1, \ldots, \lambda_n)$ and $\mathbf{b} = (\mu_1, \ldots, \mu_n)$. The expression

$$(\lambda_1 v_1 + \cdots + \lambda_n v_n) \bullet (\mu_1 v_1 + \cdots + \mu_n v_n)$$

expands, by the bilinearity of $\bullet$, to give

$$(\lambda_1 \cdot \mu_1)(v_1 \bullet v_1) + \cdots + (\lambda_i \cdot \mu_j)(v_i \bullet v_j) + \cdots + (\lambda_n \cdot \mu_n)(v_n \bullet v_n),$$

i.e. $a \bullet b$ is the sum of all vectors of the form $(\lambda_i \cdot \mu_j)(v_i \bullet v_j)$, and hence is determined by the n^2 values $(v_i \bullet v_j)$ for $1 \leq i, j \leq n$. Moreover, if G is the $n \times n$ matrix whose i-j-th entry is $(v_i \bullet v_j)$, the last equation can be written, in the language of matrices, as

$$a \bullet b = (\lambda_1 \cdots \lambda_n) G \begin{pmatrix} \mu_1 \\ \vdots \\ \mu_n \end{pmatrix},$$

which is the form of presentation of inner products used in Chapter 1. Regarding the coordinate vector itself as a row matrix, the representation may be given more succinctly as

$$a \bullet b = \mathbf{a}\, G\, \mathbf{b}^T, \qquad (\dagger)$$

(where $\mathbf{b}^T$ is the matrix *transpose* of $\mathbf{b}$).

The derivation of $(\dagger)$ used only the bilinearity of $\bullet$, but it leads to a matrix G which is symmetric, since $(v_i \bullet v_j) = (v_j \bullet v_i)$. Conversely, any $n \times n$ matrix G defines a bilinear function by the equation $(\dagger)$, and this will be symmetric if G is a symmetric matrix.

Thus inner products on a finite-dimensional vector space are completely determined by symmetric matrices, relative to chosen coordinatisations. The matrix G depends on the choice of basis for V, and so the matrix G' derived from a different basis may produce the same inner product as G (i.e. the same values for $a \bullet b$). For example, the inner product of the Lorentz plane is represented by the matrix

$$\begin{pmatrix} 0 & 1 \\ 1 & 0 \end{pmatrix},$$

relative to a suitable basis (cf. §2.5). In general, G and G' will represent the same inner product when $G' = P^T G P$, where P is the matrix transforming coordinates of the G-basis to those of the G'-basis (Snapper and Troyer [1971], Chapters 25, 26).

A *linear subspace* of vector space V is a set U of vectors that is closed under vector addition and scalar multiplication. Such a set U is a vector space itself, and contains the "origin" o of V. If $a \in V$, then the set of vectors

$$a + U = \{a + u : u \in U\}$$

is called an *affine subspace* of V with *direction space* U. $a + U$ is the affine space *through a in the direction of* (parallel to) U, and is defined to have the same dimension as that of U. Affine subspaces can also be characterised as those subsets W of V with the property that if $c, d \in W$ then the whole line cd through c and d lies in W (Birkhoff and MacLane [1965], Ch.IX).

A *hyperplane* in an n-dimensional vector space is an affine subspace of dimension $n - 1$, e.g. a line in F^2, a plane in F^3, a "threefold" in F^4, etc. In F^n, the hyperplanes are precisely the solution sets of linear equations

$$l_1 \cdot x_1 + \cdots + l_n \cdot x_n + l_{n+1} = 0$$

in n variables $x_1, \ldots, x_n$, with $l_1, \ldots, l_n$ not all zero.

The next result is well motivated by the geometric examples of Chapter 1.

Lemma 2.2.3. *Let l be a non-singular vector in an n-dimensional metric vector space. Then for any vector a, the set*

$$W = \{u \in V : l \bullet (u - a) = 0\}$$

is a hyperplane through a.

Proof. Let G be the $n \times n$ matrix representing the inner product of V relative to some coordinatisation, as in the equation (†) above. Then lG is a $1 \times n$ matrix $(l_1, \ldots, l_n)$, with at least one $l_i \neq 0$, or else lG is the zero matrix, making

$$l \bullet u = lG\, u^T = 0$$

for any vector u, contrary to the nonsingularity of vector l.

Then if u is any vector, with $u = (x_1, \ldots, x_n)$,

$$l \bullet u = lG\, u^T = l_1 \cdot x_1 + \cdots + l_n \cdot x_n.$$

Since $l \bullet (u - a) = l \bullet u - l \bullet a$, it follows that

$$l \bullet (u - a) = 0 \quad \text{iff} \quad l_1 \cdot x_1 + \cdots + l_n \cdot x_n + l_{n+1} = 0,$$

where $l_{n+1} = -(l \bullet a)$, showing that W is the solution set of a linear equation in n variables. That $a \in W$ is clear, since $l \bullet (a - a) = l \bullet o = 0$. □

Applying this to the case $n = 2$, let L be a nonsingular line in a two-dimensional vector space, with nonsingular direction vector l. For any point a, the set W of the Lemma will be a *line* through a. If u is any point of W other than a, $(u - a)$ is a direction vector for W, so the definition of W implies that $L \perp W$. But if M is any line through a that is orthogonal to L, then taking a point $u \neq a$ on M gives $(u - a)$ as a direction vector for M, hence $l \bullet (u - a) = 0$, as $L \perp M$, giving $u \in W$ and showing that $W = M =$ the unique line through a and u. This proves

Theorem 2.2.4. *In a two-dimensional vector space over a field, if L is a nonsingular line, then through each point there passes exactly one line that is orthogonal to L.* □

Full details of the theory of metric vector spaces may be found in the books of Artin [1957] and Snapper and Troyer [1971].

2.3 Metric Planes

Let α be an affine plane that carries a binary relation $\perp$ on its set of lines. If two lines L and M stand in this relation to each other, i.e. $L \perp M$, it will be said that L is *orthogonal* to M. L is *singular* in α if it is orthogonal to every line in α, and *nonsingular* otherwise. A line that is self-orthogonal, $L \perp L$, is *null*, or *isotropic*.

The structure $(\alpha, \perp)$ is a *metric plane* if it satisfies the following postulates.

O1. SYMMETRY AXIOM. $L \perp M$ *implies* $M \perp L$.

O2. ALTITUDES AXIOM. *If L is a nonsingular line, then through each point there passes exactly one line that is orthogonal to L.*

O3. QUADRUPLE AXIOM. *If a, b, c and d are four distinct points, with line ab orthogonal to line cd, and ac orthogonal to bd, then ad is orthogonal to bc.*

A *null* plane is one in which every line is singular, i.e. all lines are orthogonal to each other. A *singular* plane is one that contains at least one singular line, but is not a null plane (and so must have a nonsingular line as well). If all lines are nonsingular, $(\alpha, \perp)$ itself will be called a *nonsingular plane*. A nonsingular plane with at least one null line is an *isotropic plane*. A plane with no null lines is *anisotropic*.

It will be seen eventually that in general there is only one isotropic, and only one singular, metric geometry over any Pappian affine plane, and that these are characterised by the same inner products as in the Lorentz and Robb planes respectively. There is of course only one null plane over any α, and this is characterised by the inner product with matrix

$$\begin{pmatrix} 0 & 0 \\ 0 & 0 \end{pmatrix},$$

giving $a \bullet b = 0$ for all a and b.

Consequences of the Symmetry and Altitudes axioms

The unique line specified in O2 is called the *altitude to L through a* (note: a could lie on L), and is illustrated as the line M in Figure 2.16(i), along with the other altitudes of L, which form a class of parallel lines.

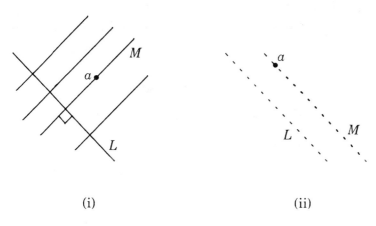

(i)   (ii)

Figure 2.16

The diagram is accurate in the case of the Euclidean plane, but in the Lorentz plane the lines that are orthogonal to a *null* line are parallel to it (and null themselves). In that case the situation is as in Figure 2.16(ii). These descriptions apply to all planes, and follow from the first two of our axioms.

Theorem 2.3.1. *If L is a nonsingular line, and $L \perp M$, then for any line N,*

$$L \perp N \quad \text{if and only if} \quad M \parallel N.$$

Proof. Suppose that $L \perp N$. If $M = N$, then immediately $M \parallel N$, so we may assume that $M \neq N$. But then if M were not parallel to N, the two lines would meet at a point, say a (Figure 2.17). M and N would then be *distinct* altitudes to L through a, contrary to O2.

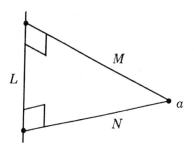

Figure 2.17

Conversely, suppose $M \parallel N$, and take a point a on N. Let K be the altitude to L through a (Figure 2.18). Since $L \perp K$, the first part of the proof shows that $M \parallel K$. Since $M \parallel N$, this yields $K \parallel N$. But K and N have a point, a, in common, so they are the same line, giving $L \perp N$. □

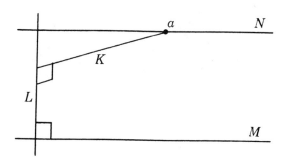

Figure 2.18

Corollary 2.3.2. *If M and N are parallel, then M is nonsingular if, and only if, N is nonsingular.*

Proof. Suppose N is nonsingular. Then there is a line L that is not orthogonal to N, so L is nonsingular also. But then if $L \perp M$, the Theorem would give $L \perp N$, since $M \parallel N$. Thus M is not orthogonal to L, and so is nonsingular. The argument is symmetric in M and N.

Corollary 2.3.3. *If L is a nonsingular line and M is one of its altitudes, then M is parallel to L if, and only if, L is a null line.*

Proof. Since $L \perp M$, putting $N = L$ in Theorem 2.3.1 gives $L \perp L$ iff $M \parallel L$.

Corollary 2.3.4. *If L is a nonsingular null line and M is parallel to L, then L is orthogonal to M, and M is null.*

Proof. Since $L \perp L$ and $L \parallel M$, replacing M by L, and N by M in Theorem 2.3.1 gives $L \perp M$. Hence $M \perp L$ and $L \parallel M$. But then, as 2.3.2 ensures M is nonsingular, the Theorem gives $M \perp M$.   □

In summary, it follows from Theorem 2.3.1 that the set $\{M : L \perp M\}$ of altitudes to a nonsingular line L form a parallelism class (which is nonempty by O2). Moreover, from the Corollaries it follows that

(a) if L is null, L is orthogonal to precisely the lines parallel to L, all of which are null (Figure 2.16(ii)), and

(b) if L is not null, the lines orthogonal to L all intersect L and are parallel to each other (Figure 2.16(i)).

The case of singular lines will be taken up in §2.4.

Consequences of the Quadruple axiom
The axiom O3 could be weakened by strengthening its hypothesis to require that no three of the four points a, b, c, d are collinear. Figure 2.19 illustrates what occurs if, say, a, b, c are collinear, with $ab \perp cd$ and $ac \perp bd$.

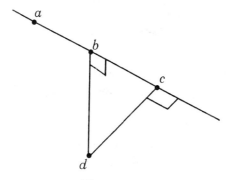

Figure 2.19

If d lies on ab then the desired conclusion $ad \perp bc$ reduces to the hypothesis $ab \perp cd$ - there is just one null line. If $ab \ (= bc)$ is a singular line, then by definition we have $bc \perp ad$, and hence the desired conclusion by the Symmetry axiom. But there are in fact no other possibilities, for if ab were nonsingular and d did not lie on ab, then cd and bd would be distinct altitudes (since $b \neq c$) to ab through d, contrary to O2.

The next point to note is that the Quadruple axiom is redundant in a singular plane, being, as will be seen in §2.4, derivable from the other two. For the rest of this section then, suppose that we are dealing with a *nonsingular* plane. In the Euclidean plane, O3 typically takes the form of Figure 2.20,

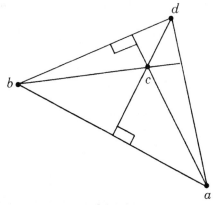

Figure 2.20

while in the Lorentz plane there are other possibilities, illustrated in Figure 2.21, since orthogonal lines can be parallel (if null).

Figure 2.21(iii) points out an important fact, namely that in the Lorentz plane the diagonals of any parallelogram of null lines are always orthogonal. In Euclidean geometry, a parallelogram only has orthogonal diagonals if it is a rhombus (all sides of equal length).

Figure 2.20 recalls another property that is familiar from Euclidean plane geometry. Since it is being assumed that all lines are nonsingular, it follows that through each vertex of any given triangle there is a unique line orthogonal to the opposite side to that vertex. These three lines are the altitudes of the triangle, and the Euclidean plane has the

ORTHOCENTRE PROPERTY: *the altitudes of a triangle are concurrent*

(the point of concurrence is the *orthocentre* of the triangle).

This property holds also in the Lorentz plane, although there the situation may look somewhat different, in that an altitude can be parallel to the opposite side, as in Figure 2.21. In each of the diagrams of that Figure, c is the orthocentre of triangle abd.

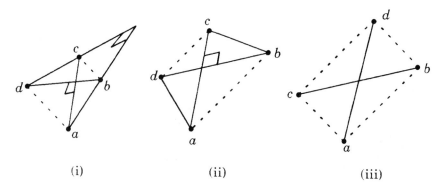

(i) (ii) (iii)

Figure 2.21

Theorem 2.3.5. *In any structure* $(\alpha, \perp)$ *that satisfies the Symmetry and Altitudes axioms and has no singular lines, the Quadruple axiom is equivalent to the Orthocentre property.*

Proof. The proof is a little more delicate than might at first appear. Suppose that a, b, d are the distinct non-collinear vertices of a triangle in α. Let L be the altitude through a, so that $bd \perp L$, and M the altitude through d, so that $ab \perp M$. Then in fact L and M must intersect. For, since $bd \perp L$, if $L \parallel M$ it would follow that $bd \perp M$ (Theorem 2.3.1). But $ab \perp M$, and so this would mean that ab and bd were distinct altitudes to M through b.

So, let L and M intersect at c. If c is distinct from a, b and d, we have the situation of Figure 2.20, and the Quadruple axiom implies $ad \perp bc$. Hence bc is the third uniquely determined altitude of the triangle, and all three of them concur at c. If however c is one of the vertices, then, depending on whether $a \neq c$ or $a = c$, we find that at least one of the sides ac and dc is an altitude. In other words, the two sides meeting at c are orthogonal (Figure 2.22) and so form the altitudes that pass through the other two vertices. Again all three altitudes concur at c.

It is left as an exercise for the reader to, conversely, derive O3 from the Orthocentre property. The proof has been almost given in the above discussion. □

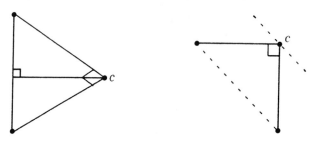

Figure 2.22

Thus far, the affine plane α has not been assumed to satisfy any special incidence properties, but in fact the orthogonality axioms impose strong restrictions here. This is shown by the following construction, discovered by Schur [1903].

Theorem 2.3.6. *In a nonsingular metric plane, any two orthogonal lines have the Pappus property.*

Proof. Let L and L' be distinct orthogonal lines, and let a, b, c and a', b', c' be triples of distinct points on L and L', respectively, such that $ab' \parallel a'b$ and $ac' \parallel a'c$. It has to be shown that $bc' \parallel b'c$.

Recall from §2.1 that if any point of one triple is collinear with the other triple, then $L = L'$, contrary to the present hypothesis.

Let M be the altitude to ac' through b. Then M must intersect L': for otherwise, if $M \parallel L'$, then since $ac' \perp M$, it would follow that $ac' \perp L'$ (2.3.1) and hence, since $L \perp L'$, that L and ac' would be altitudes to L' through a that are distinct, as c' is not on L, contradicting axiom O2. (Note that all of this is compatible with having $L \parallel L'$.)

Thus let M meet L' at d, and consider the resulting configuration as in Figure 2.23. For each of the points numbered 1 - 7 in that diagram, the pair of lines determining the point prove to be orthogonal. In particular, points 6 and 7 indicate that $b'c$ and bc' are both altitudes to ad, and hence are parallel, as desired.

Therefore, to prove the Theorem it suffices to establish the orthogonality relations numbered 1 - 7, and this is done as follows.

1: $L \perp L'$ holds by hypothesis.

2: $bd \perp ac'$ holds by the definition of $M = bd$.

3: $bd \perp ac'$ (by 2) and $ac' \parallel a'c$ (by hypothesis) imply $bd \perp a'c$.

4: if $d = a'$, then the result $bd \perp a'c$ (3) can be rewritten as $ba' \perp dc$, which is the desired conclusion. But if $d \neq a'$, then

c, d, a' form a triangle with altitudes bc (1) and bd (3) meeting at b. Hence by the Orthocentre property, ba' is the third altitude, giving $ba' \perp dc$.

5: $a'b \perp dc$ (4) and $ab' \parallel a'b$ (hypothesis) imply $ab' \perp dc$.

6: if $d = b'$, then $ab' \perp dc$ (5) becomes the desired conclusion $ad \perp b'c$. Otherwise, the Orthocentre property applies to triangle cdb': altitudes ac (1) and ab' (5) meet at a, so the third altitude is ad, giving $ad \perp b'c$.

7: if $d = c'$, the desired conclusion $ad \perp bc'$ simply restates the original construction $ac' \perp M$. Otherwise, triangle bdc' has altitudes ab (1) and ac' (2), and hence altitude ad, giving $ad \perp bc'$. $\qquad \square$

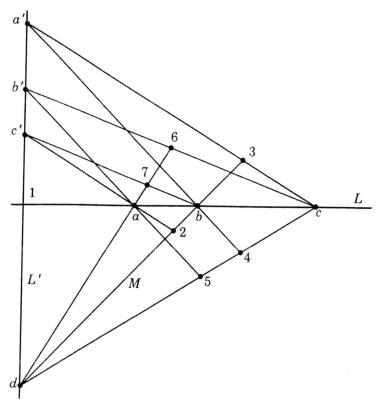

Figure 2.23

Observe that in Theorem 2.3.6 it is quite possible for the orthogonal lines L and L' to be parallel (if null). But in §2.1 it was noted that for a Desarguesian plane to be Pappian it suffices that the Pappus property hold for at least one pair of *intersecting* lines:

Corollary 2.3.7. *If a nonsingular Desarguesian metric plane has a pair of intersecting orthogonal lines, then it is Pappian.*   □

This result becomes particularly significant in affine spaces of three or more dimensions, where the Desargues property can be *proven* from the axioms of an affine space. In such spaces, the Pappus property will follow from the appropriate orthogonality axioms alone (cf. Chapter 4).

The hypothesis of nonsingularity is unavoidable in Corollary 2.3.7, for the universal orthogonality relation that makes all lines singular satisfies in any Desarguesian plane, Pappian or not, all hypotheses of the Corollary except nonsingularity. Also, in §2.5 there will be considered a pathological situation involving Fano planes which shows that there can be non-Pappian examples satisfying all hypotheses of the Corollary except for the existence of *intersecting* orthogonal lines.

Isomorphism of Metric Planes

Each species of mathematical structure has its own notion of "sameness", or isomorphism, of members of that species. It is possible for two metric planes to have different (non-isomorphic) orthogonality structures while the affine planes on which they are based are isomorphic. For instance, the Euclidean and Lorentz planes: their underlying affine parts are identical.

A pair of metric geometries $(\alpha, \perp)$ and $(\alpha', \perp')$ are *isomorphic* if there is an isomorphism f of α onto α' (§2.1) that has

$$L \perp M \quad \text{if and only if} \quad f(L) \perp f(M)$$

for all α-lines L, M, where $f(L)$ is the unique line incident with all points of $\{f(a) : a\mathcal{I}L\}$.

Now suppose that α_F and $\alpha_{F'}$ are metric field-planes, and that their orthogonality relations $\perp$ and $\perp'$ are given by inner products, i.e. in α_F it holds that

$$\mathbf{o}a \perp \mathbf{o}b \quad \text{if and only if} \quad a \bullet b = 0,$$

and similarly for $\perp'$ in $\alpha_{F'}$. Furthermore, suppose that f is an isomorphism between α_F and $\alpha_{F'}$ that takes the origin in α_F to the origin in $\alpha_{F'}$. Then it can be seen that f will be an isomorphism from $(\alpha_F, \perp)$ onto $(\alpha_{F'}, \perp')$ just in case

$$a \bullet b = 0 \quad \text{if and only if} \quad f(a) \bullet' f(b) = 0$$

for all points a, b in α.

2.4 The Singular Plane

In the singular Robb plane, the notion of an altitude of a triangle does in fact make sense, since through each point there is only one line that is orthogonal to the opposite side - the one parallel to the y-axis. A triangle will have two or three distinct (and singular) altitudes, depending on whether one or none of its sides are singular (Figure 2.24).

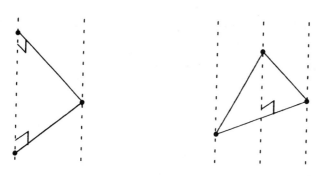

Figure 2.24

These singular altitudes are all parallel and may be regarded as meeting at a point "at infinity" (cf. §3.3). Thus the Robb plane satisfies a generalised version of the Orthocentre property. In this section it will be seen that the description holds good for all singular planes, because over any affine plane there is essentially only one such orthogonality structure - the one characterised by the same inner product that the Robb plane has.

Theorem 2.4.1. *In any metric plane, if L is singular and M is parallel to L, then M is also singular.*

Proof. This is a restatement of Corollary 2.3.2.

Theorem 2.4.2. *In a singular plane, there is exactly one singular line through each point, all singular lines are parallel, and all lines are orthogonal to these, and only these, singular lines.*

Proof. Let L be a singular line. Since the plane is non-null, it contains a non-singular line M. By the previous theorem, L and M cannot be parallel, so they meet at some point, say a (Figure 2.25).

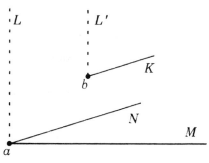

Figure 2.25

Since M is nonsingular, the Altitudes axiom implies that L is the only line through a orthogonal to M. Thus if N is any other line through a, N is not orthogonal to M, hence N is nonsingular, and the same argument makes L the unique altitude to N through a.

This establishes that in the pencil of all lines through a, L is singular and all other lines are orthogonal only to L. But the parallelism properties derived in §2.3 then allow this description to be transferred to the pencil of lines at any other point b, and to thereby complete the theorem. For, the line L' through b parallel to L will be singular (2.4.1), and any other line K through b will be parallel to some line N through a with $N \neq L$. Then nonsingularity of N implies that K is nonsingular and has L' as its unique altitude through b. In particular, this shows that all lines not parallel to L are nonsingular, so the singular lines are *precisely* those parallel to L.

Theorem 2.4.3. (Criteria for Nullity of a Plane). *A metric affine plane is a null plane if it satisfies any one of the following conditions.*
(1). *There exist two intersecting singular lines.*
(2). *There exists a singular line intersected by a null line.*
(3). *There exist two intersecting and orthogonal null lines.*

Proof.
    (1). If the plane is non-null and contains singular lines (i.e. is a singular plane), then these lines are all parallel (Theorem 2.4.2).
    (2). If L is a singular line, M intersects L at a, say, and the plane is not null, then M is orthogonal only to L through a, hence is not orthogonal to itself, i.e. is not a null line.
    (3). A nonsingular null line is orthogonal only to lines parallel to it (Corollary 2.3.3). Hence a pair of intersecting null lines must

both be singular, making the plane null by part 1 of this Theorem.

Coordinatising a Singular Plane

Let $(\alpha, \perp)$ be a Pappian singular plane. Theorem 2.4.2 gives a complete description of the structure of the relation $\perp$. There is at least one singular line L, all other singular lines are parallel to L, and all lines parallel to L are singular. Thus the set of singular lines is precisely the parallelism class $\{M : L \parallel M\}$ of L, and all lines in α are orthogonal to just these singular lines.

The analysis is identical to that of the Robb plane, and gives a solution to the coordinatisation problem: in constructing coordinates for α, take the singular line L as the y-axis, and introduce the inner product with matrix

$$\begin{pmatrix} 1 & 0 \\ 0 & 0 \end{pmatrix},$$

which makes two vectors orthogonal if, and only if, at least one of them lies on the y-axis. This reflects the structure of $(\alpha, \perp)$ exactly.

In sum then, it has now been established that the same coordinatisation can be made to work for any given singular geometry over a Pappian affine plane α_F, i.e. that there is essentially only one such geometry, which will be called the *singular plane over F*.

The attentive reader will have noticed that in reaching this conclusion no use has been made of the Quadruple axiom at all. It follows that if a structure $(\alpha, \perp)$ has a singular line and a nonsingular line, and satisfies O1 and O2, then it must satisfy O3. The proof is an easy consequence of Theorem 2.4.2. For, if both $ab \perp cd$ and $ac \perp bd$, then in each case at least one of the pair of orthogonal lines must be singular. Take the case that ab and bd are singular. Then $ab \parallel bd$ and so $ab = bd$. Thus d lies on ab, so that $ad \ (= ab)$ is singular, giving the desired conclusion $ad \perp bc$ for O3. The other three cases are similar.

The analysis of the basic orthogonality relations between both singular and nonsingular lines, as encapsulated in the Symmetry and Altitude axioms, is now complete. Given any two points a and b in a metric plane, the $\perp$-structures of the two pencils of lines determined by a and b "look the same". The function f that assigns to each line through a the unique line $f(L)$ through b that is parallel to L establishes a one-to-one correspondence between the members of these pencils that has

$$L \perp M \quad \text{if and only if} \quad f(L) \perp f(M),$$

no matter what type of line L or M is.

2.5 The Artinian Plane

In two-dimensional spacetime, the two null lines of slope $+1$ and -1 are perpendicular in the Euclidean sense, but not orthogonal in the sense of the Lorentzian inner product

$$\begin{pmatrix} 1 & 0 \\ 0 & -1 \end{pmatrix}.$$

The non-null x and t axes are on the other hand orthogonal in both senses. We could however reverse this situation by coordinatising the plane so that the two null lines become the coordinate axes. The easiest way to visualise this is to imagine the plane being rotated clockwise about the origin through an angle of 45°, so that the null line of slope -1 becomes the vertical axis, and the other one the horizontal axis (Figure 2.26).

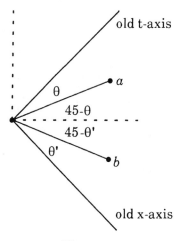

Figure 2.26

According to the account of Chapter 1, the lines oa and ob as shown are orthogonal in the spacetime sense if, and only if, the angle that oa makes to the old t-axis is the same as that which ob makes to the old x-axis, i.e. $\theta = \theta'$. But then this is equivalent to having oa and ob both at angles $45 - \theta$ to the horizontal null line, i.e. that *relative to the new null axes*, the slope of ob is the negative of the slope of oa. If a has coordinates (x_1, y_1), and b has (x_2, y_2), relative to the new axes, then the slope of oa will be y_1/x_1, that of ob will be y_2/x_2, and so

$$oa \perp ob \quad \text{iff} \quad y_1/x_1 = -y_2/x_2$$
$$\text{iff} \quad x_2 y_1 = -x_1 y_2.$$

Putting

$$a \bullet b = x_2 y_1 + x_1 y_2,$$

gives

$$oa \perp ob \quad \text{iff} \quad a \bullet b = 0.$$

Since the inner product $x_2 y_1 + x_1 y_2$ makes the vectors $(1,0)$ and $(0,1)$ self-orthogonal, the result is another characterisation of the geometry of two-dimensional spacetime, this time by the inner product with matrix

$$\begin{pmatrix} 0 & 1 \\ 1 & 0 \end{pmatrix}.$$

It will now be shown that this works in general for isotropic planes.

A parallelogram $abcd$ in a metric plane is called a *null parallelogram* if its four sides are all null lines (Figure 2.27).

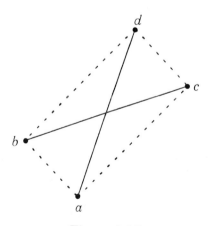

Figure 2.27

Such figures in spacetime are called *optical parallelograms* in Robb [1914]. They cannot occur of course in a singular plane, where null lines do not intersect.

Theorem 2.5.1. *In a nonsingular plane, the diagonals of a null parallelogram are orthogonal.*

Proof. (cf. Figure 2.27). By Corollary 2.3.4, the opposite sides of a null parallelogram are orthogonal (being null and parallel) in a nonsingular plane. Hence $ab \perp cd$ and $ac \perp bd$, from which the desired conclusion $ad \perp bc$ is immediate by the Quadruple axiom.

□

From now on, until the end of this section, assume that $(\alpha, \perp)$ is a *Pappian* plane that is *isotropic* (i.e. is nonsingular and has a null line), and is not a Fano plane. Thus the diagonals of any parallelogram in α always intersect.

Theorem 2.5.2. *Through any point in α there passes at most two null lines.*

Proof. Suppose that three distinct null lines L, M, N pass through point a.

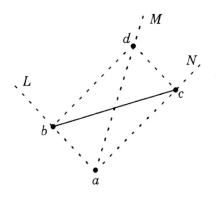

Figure 2.28

Take a point d on M that is distinct from a. Then the line through d parallel to N intersects L at a point b distinct from a. Similarly the line through d parallel to L meets N at a point $c \neq a$. Since lines parallel to null lines are null (2.3.4), $abcd$ is a null parallelogram, and so $ad \perp bc$. But then as ad is null, it follows that $ad \parallel bc$ (2.3.3), contrary to the assumption that diagonals intersect.

Theorem 2.5.3. *There exists a pair of intersecting null lines.*

Proof. As the plane is isotropic, there exists a null line L. Take a point o on L and let M be a line through o distinct from L. Let N be the altitude to M through o. If $M = N$, then M is null and the proof is finished. Otherwise, take a point p on L distinct from o and let the lines through p parallel to M and N meet N and M, respectively, at a and d, so that $apod$ is a parallelogram (Figure 2.29).

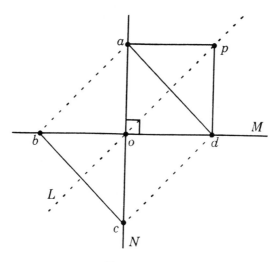

Figure 2.29

As α is not a Fano plane, the diagonal ad must intersect the null diagonal op. The proof from here consists in showing that ad is itself null.

So, let the lines through a and d that are parallel to L meet M and N, respectively, at b and c. Since L is null, so too are ab and cd. But then ab and cd are null and parallel, making $ab \perp cd$. The original construction gave $ac \perp bd$. Thus by the Quadruple axiom, $ad \perp bc$. But applying the Little Desargues property to the triangles apd and boc, which may be done since $ab \parallel po \parallel dc$, we have $ap \parallel bo$ and $pd \parallel oc$, yielding $ad \parallel bc$. Thus ad has a parallel altitude in bc, and so must be a null line.

Corollary 2.5.4. *Through any point in α there passes exactly two null lines.*

Proof. There exists a pair of intersecting null lines, say L and L'. Then for any point a, the lines through a parallel to L and L' are null, and there can be no others by 2.5.2.   □

In coordinatising the affine part of our isotropic plane, we may thus take a pair of intersecting null lines as coordinate axes (Figure 2.30). Moreover, as the plane is not null these axes are not orthogonal (Theorem 2.4.3).

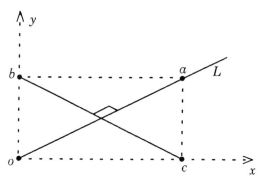

Figure 2.30

Let L be any (necessarily non-null) line through the origin o that is distinct from the two null axes. If a is any point on L distinct from o, the lines through a parallel to the axes are null, forming the parallelogram $obac$ as shown. Theorem 2.5.1 then gives $oa \perp bc$. As explained with the coordinatisation procedure described in §2.1, if a has coordinates (x_1, y_1), then $b = (0, y_1)$ and $c = (x_1, 0)$. Also the line oa has slope y_1/x_1, while bc has slope

$$\frac{0 - y_1}{x_1 - 0} = -\frac{y_1}{x_1}.$$

Since parallel lines are those with the same slope, this establishes that the slope of any altitude to a non-null line is the negative of the slope of the line.

Now let $a = (x_1, y_1)$ and $b = (x_2, y_2)$ be any two points on neither axis (Figure 2.31). If oa has slope m, while ob has slope n, then $y_1 = m \cdot x_1$ and $y_2 = n \cdot x_2$. Moreover $x_1 \cdot x_2 \neq 0$, since neither $x_1 = 0$ nor $x_2 = 0$. Thus

$$
\begin{aligned}
oa \perp ob \quad &\text{iff} \quad m = -n \\
&\text{iff} \quad m + n = 0 \\
&\text{iff} \quad x_1 \cdot x_2 \cdot (m + n) = 0 \qquad \text{since } (x_1 \cdot x_2 \neq 0) \\
&\text{iff} \quad x_2 \cdot x_1 \cdot m + x_1 \cdot x_2 \cdot n = 0 \\
&\text{iff} \quad x_2 \cdot y_1 + x_1 \cdot y_2 = 0,
\end{aligned}
$$

and so the inner product

$$\begin{pmatrix} 0 & 1 \\ 1 & 0 \end{pmatrix}$$

characterises the ⊥-structure of our isotropic plane.

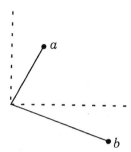

Figure 2.31

The situation reached here is parallel to that for the singular plane in the previous section. From the knowledge that $(\alpha, \perp)$ is a Pappian isotropic non-Fano plane, a coordinatisation of it has been completely determined. Thus there is essentially only one isotropic geometry over any Pappian field-plane α_F when the characteristic of F is not 2. This plane has been dubbed (by Snapper and Troyer [1971]) the *Artinian* plane over F, after Emil Artin, whose book *Geometric Algebra* [1957] is a classic in the study of metric vector spaces.

The Artinian plane over the real affine plane is of course the same geometrical structure as the Lorentz plane, but the latter name will be reserved for the coordinatisation with inner product

$$\begin{pmatrix} 1 & 0 \\ 0 & -1 \end{pmatrix}.$$

This matrix defines an isotropic plane over any field F, and if the characteristic of F is not 2 then this will be the Artinian plane over F, and hence can be recoordinatised as above. The two coordinatisations may be seen as two different ways of labelling the points of the same geometrical structure (alias), or as representing two isomorphic metric planes (alibi). It is left to the reader to give an explicit description of a collineation that establishes this isomorphism.

Exercises.
Let $(\alpha, \perp)$ be an isotropic non-Fano plane.
  1. Extract fron the proof of Theorem 2.5.3 a proof that if a parallelogram has adjacent sides orthogonal, and one diagonal is null, then the other diagonal is also null.

2. Use Exercise 1 to show that α may be coordinatised with orthogonal non-null axes so that the two null lines through the origin have equations $y = x$ and $y = -x$:

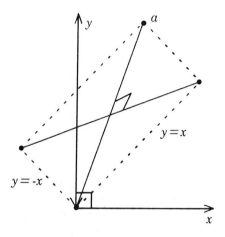

Figure 2.32

3. With the coordinatisation of Figure 2.32, if $a = (x_1, y_1)$, so that the slope of oa is y_1/x_1, show that any altitude to oa has slope x_1/y_1. Hence give a direct coordinatisation of $(\alpha, \perp)$ in terms of the inner product

$$\begin{pmatrix} 1 & 0 \\ 0 & -1 \end{pmatrix}.$$

□

If a nonsingular metric plane contains a non-null line L, then it must contain a pair of intersecting orthogonal lines: any altitude of L will intersect L. But since any point in an affine plane has at least three lines passing through it, Theorem 2.5.2 implies that a non-Fano plane always has a non-null line, and hence a pair of intersecting orthogonal lines (and so by Corollary 2.3.7 is Pappian if Desarguesian). But what if α is a Fano plane? Here it is possible, in just one case, for all lines to be null. If so, there are no intersecting orthogonal pairs (Theorem 2.4.3(3)), and the relation $\perp$ is simply the relation of *parallelism*. The point is that the latter satisfies O1, O2, and O3 in (and only in) a Fano plane. Symmetry is obvious, the Altitude axiom becomes the axiom of Parallels (A3), and the Quadruple axiom becomes the statement that the diagonals of a parallelogram are parallel. We shall call this the *degenerate* metric

Fano plane. It can be characterised by the inner product

$$\begin{pmatrix} 0 & 1 \\ 1 & 0 \end{pmatrix}$$

that has been used for isotropic planes in the presence of Fano's axiom. In the absence of the latter, the field satisfies $x = -x$, and so using this inner product gives

$$
\begin{aligned}
oa \perp ob \quad &\text{iff} \quad x_2 \cdot y_1 + x_1 \cdot y_2 = 0 \\
&\text{iff} \quad \frac{y_1}{x_1} = -\frac{y_2}{x_2} = \frac{y_2}{x_2} \\
&\text{iff} \quad \text{slope } oa = \text{slope } ob \\
&\text{iff} \quad oa \parallel ob \quad (\text{i.e. } oa = ob).
\end{aligned}
$$

The inner product

$$\begin{pmatrix} 1 & 0 \\ 0 & -1 \end{pmatrix} = \begin{pmatrix} 1 & 0 \\ 0 & 1 \end{pmatrix}$$

produces a second isotropic Fano plane over a field in which $1 = -1$. This will be referred to as an *Artinian Fano plane*. It has exactly one null line through the origin: the one with equation $y = x$, or equivalently $y = -x$. The classification of isotropic Fano planes will be completed in the next section, where it will be shown (Theorem 2.6.7) that the two types just described are all that there are.

2.6 Constants of Orthogonality

In the Lorentz plane, when two lines are orthogonal their slopes are reciprocal to each other (this includes the null lines, but not of course vertical lines as they are not assigned a slope). Thus the product of two such slopes is always $+1$. An analogous situation obtains in the Euclidean plane, where the product of the slopes of perpendicular lines is always -1. Each of these planes has then a special number k associated with it, called its *constant of orthogonality*. Furthermore, in each case the $\perp$-structure of the plane is characterised by the inner product

$$a \bullet b = -kx_1x_2 + y_1y_2,$$

defined by the matrix

$$\begin{pmatrix} -k & 0 \\ 0 & 1 \end{pmatrix},$$

since in the Euclidean case $-k = +1$, while the Lorentz plane ($k = +1$) has

$$-x_1x_2 + y_1y_2 = 0 \quad \text{iff} \quad x_1x_2 - y_1y_2 = 0.$$

Notice that the nature of the constant k is dependent on the choice of coordinates for the plane. In the real Artinian plane, characterised by

$$\begin{pmatrix} 0 & 1 \\ 1 & 0 \end{pmatrix},$$

the product of orthogonal slopes is -1, and this constant does *not* give the inner product as above. The link between constant of orthogonality and inner product holds only for coordinatisations with orthogonal axes. In the present section it will be shown that every nonsingular metric plane, except for the degenerate Fano case, has a constant of orthogonality and can be characterised by the associated inner product $-k \cdot x_1 \cdot x_2 + y_1 \cdot y_2$. This will give an alternative construction for the isotropic plane, and also cover the non-degenerate Fano planes.

Suppose now that $(\alpha, \perp)$ is any nonsingular Pappian metric plane that is not a degenerate Fano plane (i.e. $\perp$ is not the parallelism relation). At the end of the last section it was observed that α must contain a pair of intersecting orthogonal lines. Let the plane be coordinatised with these as axes (Figure 2.33).

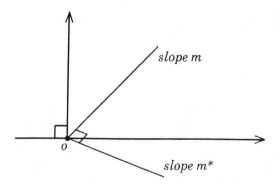

Figure 2.33

Then if m is any non-zero element of the coordinatising field F for α, the line L through the origin of slope m is distinct from both axes. Hence its altitude M through o is distinct from both axes (by O2). Therefore M has a slope m^* which is non-zero. The following construction, due to Baer [1944], yields the constant of orthogonality.

Theorem 2.6.1. *For any two non-zero elements* m *and* n *of* F, $m \cdot m^* = n \cdot n^*$.

Proof.

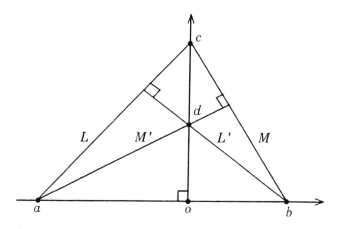

Figure 2.34

Let L be the line through the point $a = (n, 0)$ of slope m. Then L has an equation of the form

$$y = m \cdot x + j.$$

Since L contains the point $(n, 0)$, the equation implies that $0 = m \cdot n + j$, and so the equation of L is

$$y = m \cdot x - m \cdot n.$$

Similarly, the line M through $b = (m, 0)$ of slope n has equation

$$y = n \cdot x - n \cdot m.$$

Since $m \neq n$, L and M are not parallel and so meet at a point, say c, whose coordinates (x, y) satify the equations of both lines. This requires that

$$m \cdot x - m \cdot n = n \cdot x - n \cdot m,$$

and hence

$$m \cdot x = n \cdot x$$

(since $m \cdot n = n \cdot m$ by Pappus!).

But, again since $m \neq n$, this last equation can only be true if $x = 0$. Thus c has coordinates $(0, -m \cdot n)$ and lies on the y-axis.

Now the latter was constructed to be orthogonal to the x-axis, i.e. to ab, and so is the altitude to triangle abc through c. Thus by the Orthocentre property, the other two altitudes L' and M' meet at a point d on the y-axis, i.e. on oc as shown in Figure 2.34.

Now L' has slope m^*, since all lines orthogonal to L are parallel and have therefore the same slope. Since L' passes through $b = (m, 0)$ it follows that it has equation

$$y = m^* \cdot x - m^* \cdot m.$$

Similarly, M' has equation

$$y = n^* \cdot x - n^* \cdot n.$$

But the coordinates of d satisfy both of these equations and, since the x-coordinate of d is 0, this implies $m^* \cdot m = n^* \cdot n$ as desired.

□

Now let k be the common value of all the products $m \cdot m^*$ for $m \neq 0$. If $a = (x_1, y_1)$ and $b = (x_2, y_2)$ are points on neither axis, with the slope of oa being $m \neq 0$ and the slope of ob being $n \neq 0$, then $y_1 = m \cdot x_1$ and $y_2 = n \cdot x_2$,

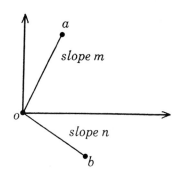

Figure 2.35

and so,

$$\begin{aligned}
oa \perp ob \quad &\text{iff} \quad m^* = n \\
&\text{iff} \quad k = m \cdot n \\
&\text{iff} \quad -k + m \cdot n = 0 \\
&\text{iff} \quad x_1 \cdot x_2 \cdot (-k + m \cdot n) = 0 \qquad (\text{since } x_1 \cdot x_2 \neq 0) \\
&\text{iff} \quad -k \cdot x_1 \cdot x_2 + m \cdot x_1 \cdot n \cdot x_2 = 0 \\
&\text{iff} \quad -k \cdot x_1 \cdot x_2 + y_1 \cdot y_2 = 0.
\end{aligned}$$

The matrix

$$\begin{pmatrix} -k & 0 \\ 0 & 1 \end{pmatrix}$$

then characterises the $\perp$-structure of the plane (this inner product gives orthogonal coordinate axes, as was arranged from the outset). The procedure however does more than just coordinatise the geometry in question. The description of the inner product in terms of an orthogonality constant gives an effective tool for determining the possible types of plane that can occur. In general, if j is any non-zero element of field F, we denote by $\alpha_F(j)$ the (necessarily nonsingular) metric plane over F with inner product

$$j \cdot x_1 \cdot x_2 + y_1 \cdot y_2,$$

defined by the matrix

$$\begin{pmatrix} j & 0 \\ 0 & 1 \end{pmatrix}.$$

Thus the plane with orthogonality constant k is $\alpha_F(-k)$, and every nonsingular metric plane over F has this form for some $k \neq 0$. Different constants may however determine the same geometry:

Theorem 2.6.2. *If there exists an $m \in F$ such that $j = l \cdot m^2$, then $\alpha_F(j)$ is isomorphic to $\alpha_F(l)$.*

Proof. Let f be the collineation

$$x' = m \cdot x$$
$$y' = y$$

that maps $a = (x_1, y_1)$ to $f(a) = (m \cdot x_1, y_1)$. Since

$$j \cdot x_1 \cdot x_2 = l \cdot m^2 \cdot x_1 \cdot x_2 = l \cdot (m \cdot x_1) \cdot (m \cdot x_2),$$

it follows that

$$a \bullet b \text{ in } \alpha_F(j) = f(a) \bullet f(b) \text{ in } \alpha_F(l),$$

so

$$a \bullet b = 0 \quad \text{iff} \quad f(a) \bullet f(b) = 0,$$

as required to make f an isomorphism of metric planes (cf. §2.3).

$\square$

In our present notation, the Artinian plane over F is $\alpha_F(-1)$ (including the Fano plane case), while $\alpha_F(1)$ is the plane with inner product

$$\begin{pmatrix} 1 & 0 \\ 0 & 1 \end{pmatrix}.$$

In the case of the real number field **R**, the latter plane is anisotropic, but in other cases it can be isotropic! To describe just when this occurs, the expression "$\sqrt{k}$ exists in F" will be used to mean that there is some $m \in F$ with $m^2 = k$ (so that m is a "square root" of k). Then for $k \neq 0$, the result is:

Theorem 2.6.3. $\alpha_F(-k)$ *is the Artinian plane over F if, and only if, $\sqrt{k}$ exists in F.*

Proof. If $\alpha_F(-k)$ is isotropic it contains a null line L, which cannot be either axis since the axes intersect and are orthogonal. Hence L has non-zero slope, m say. But all altitudes to L are parallel to L and so also have slope m. Thus $k = m \cdot m^* = m \cdot m = m^2$, making m a square root of k.

Conversely, if $k = m^2$, then $m \cdot m^* = m \cdot m$, so $m^* = m$ and lines of slope m are self-orthogonal, making $\alpha_F(-k)$ isotropic (alternatively, observe that $-k = -1 \cdot m^2$, so by Theorem 2.6.2, $\alpha_F(-k)$ is isomorphic to the isotropic plane $\alpha_F(-1)$).

Corollary 2.6.4. $\alpha_F(-k)$ *is anisotropic if, and only if, $\sqrt{k}$ does not exist in F.*   □

Thus it follows that for a field in which $\sqrt{-1}$ exists, e.g. the field **C** of complex numbers, $\alpha_F(1)$ is Artinian. **C** is an example of an *algebraically closed* field, in which all polynomial equations have solutions, and so, in particular, all elements have square roots. If F is such a field, there is only one nonsingular metric plane over F - the isotropic one.

A parallelogram in a metric plane is called a *square* if its diagonals are orthogonal and it has a pair of adjacent sides that are orthogonal (hence the other three pairs of adjacent sides will be orthogonal). The plane $\alpha_F(1)$ always contains at least one square: the one shown in Figure 2.36 whose diagonals have slopes $+1$ and -1.

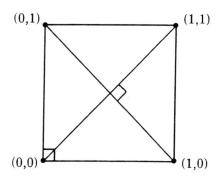

Figure 2.36

In fact the inner product

$$\begin{pmatrix} 1 & 0 \\ 0 & 1 \end{pmatrix}$$

characterises the only plane over F to contain a square.

Theorem 2.6.5. *The following are equivalent.*
(1). $\alpha_F(-k)$ *contains a square.*
(2). $\sqrt{-k}$ *exists in* F.
(3). $\alpha_F(-k)$ *is isomorphic to* $\alpha_F(1)$.

Proof. If (2) holds, then $-k = 1 \cdot m^2$ for some m, and so (3) follows by Theorem 2.6.2. On the other hand, since $\alpha_F(1)$ contains a square, any plane with identical $\perp$-structure will also, and so (3) implies (1). The burden of the proof then is to show that (1) implies (2).

So, suppose that $\alpha_F(-k)$ contains a parallelogram $abcd$ that is a square. Then we recoordinatise the plane so that two adjacent sides, say ab and ac, become the coordinate axes, with a as origin. The new coordinate system will then have orthogonal axes and so have an associated orthogonality constant k', which may be different from k. However it will follow immediately that

Lemma A. $\sqrt{-k'}$ *exists in* F.

Proof. Suppose d gets coordinates (x, y) in the new system (Figure 2.37).

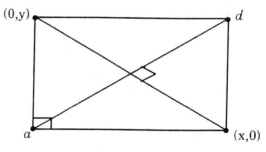

Figure 2.37

Then $b = (x, 0)$ and $c = (0, y)$. Diagonal ad has slope $m = y/x$, while bc has slope $m^* = -y/x = -m$. Thus

$$k' = m \cdot m^* = m \cdot -m = -m^2,$$

and so $-k' = m^2$, as required to prove the Lemma.

The next step will be to show

Lemma B. $k = k' \cdot n^2$ for some $n \in F$.

The two Lemmata then imply condition (2) of the Theorem, since they give

$$-k = -k' \cdot n^2 = m^2 \cdot n^2 = (m \cdot n)^2.$$

Thus to finish the proof we have to establish Lemma B, and this requires us to give the details of the proposed recoordinatisation. This is carried out in two steps, and is done from the *alias* point of view, i.e. the transformations will be construed as relabellings of the points of the plane, which themselves remain fixed.

Step 1. To give a the coordinates $(0, 0)$, perform the relabelling $(x, y) \mapsto (x', y')$, where $x' = x - p$ and $y' = y - q$ and (p, q) are the coordinates of a in the original system $\alpha_F(-k)$. This leaves the slope of any line exactly as before: a "vertical" line with equation of the form $x = l$ now has equation $x' = l - p$, i.e. one of the same form. Similarly, a line $y = l$ parallel to the old x-axis gets the new equation $y' = l - q$ and remains parallel to the new x'-axis. A line $y = m \cdot x + l$ with old slope m gets the new equation $y' + q = m \cdot (x' + p) + l$, i.e.

$$y' = m \cdot x' + (p + l - q)$$

and still has slope m. Thus the orthogonality constant is still k in the new system.

Step 2. In view of Step 1, it may be taken that $\alpha_F(-k)$ has been given with vertex a of the square as origin. Suppose that $b = (s, t)$ and $c = (u, v)$. If either ab or ac is a coordinate axis, then the other is the other axis (as $ab \perp ac$), so the square already has the form of Figure 2.37. In this case, the required k' is just k, and so $\sqrt{-k}$ exists by the proof of Lemma A.

Finally then, consider the (most involved) case that neither ac nor ab is an axis in $\alpha_F(-k)$, so that none of s, t, u, v is 0 (Figure 2.38).

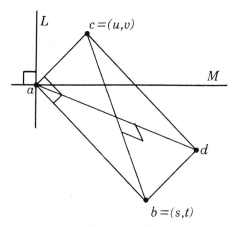

Figure 2.38

The line ac has old slope $m = v/u$, while ab has $m^* = t/s$. Since $ab \neq ac$, $m \neq m^*$, and so $\Delta \neq 0$, where

$$\Delta = s \cdot v - u \cdot t.$$

Now consider the coordinate change $(x, y) \mapsto (x', y')$, where

$$x' = \frac{v}{\Delta} \cdot x - \frac{u}{\Delta} \cdot y$$

$$y' = -\frac{t}{\Delta} \cdot x + \frac{s}{\Delta} \cdot y.$$

From the alibi point of view this is the standard linear transformation that moves b to $(1, 0)$ and c to $(0, 1)$. From the alias perspective we say, rather, that the collineation leaves a with coordinates $(0, 0)$, and gives b the new coordinates $(1, 0)$ and c the new coordinates $(0, 1)$. Thus ac must be the line with equation $x' = 0$, i.e. the new y'-axis, while ab is the new x-axis. Hence the system has the form required for Lemma A. To complete Lemma B, consider the old axes L and M as in Figure 2.38. L has equation $x = 0$, and so any point on it aquires new coordinates

$$x' = -\frac{u}{\Delta} \cdot y$$
$$y' = \frac{s}{\Delta} \cdot y$$

which satisfy

$$y' = -\frac{s}{u} \cdot x'.$$

The latter is then the new equation of L, and so L has new slope $-s/u$. Similarly, M (old equation $y = 0$) gets new equation

$$y' = -\frac{t}{v} \cdot x'.$$

But $L \perp M$, so the orthogonality constant in the new system can now be computed as

$$k' = -\frac{s}{u} \cdot -\frac{t}{v} = \frac{s}{u} \cdot \frac{t}{v}.$$

But for the old constant we had

$$k = m \cdot m^* = \frac{v}{u} \cdot \frac{t}{s},$$

so

$$\frac{k}{k'} = \frac{v}{u} \cdot \frac{t}{s} \cdot \frac{u}{s} \cdot \frac{v}{t} = \frac{v^2}{s^2},$$

and hence $k = k' \cdot n^2$, where $n = v/s$. $\square$

Taking the case $k = 1$ now gives:

Corollary 2.6.6. *The following are equivalent.*
(1). *The Artinian plane over F contains a square.*
(2). *$\sqrt{-1}$ exist in F.*
(3). *The Artinian plane can be coordinatised by the inner product* $x_1 \cdot x_2 + y_1 \cdot y_2$. $\square$

Theorem 2.6.5 allows us also to completely classify the isotropic Fano planes, showing that the two types described at the end of §2.5 cover all cases.

Theorem 2.6.7. *Suppose F has characteristic 2 and $(\alpha_F, \perp)$ is isotropic. Then either*

(1) there exists a pair of intersecting orthogonal lines, and $\perp$ is characterised by the inner product

$$\begin{pmatrix} 1 & 0 \\ 0 & 1 \end{pmatrix},$$

i.e. the geometry is the Artinian plane over F; or else

(2) $(\alpha_F, \perp)$ is the degenerate Fano plane over F, with inner product

$$\begin{pmatrix} 0 & 1 \\ 1 & 0 \end{pmatrix},$$

and $L \perp M$ if and only if $L \parallel M$.

Proof.

(1). If there exist lines L and M, with $L \perp M$ and L and M intersecting at a point a, say, then neither L nor M can be null (Corollary 2.3.3). But since the plane is isotropic, there must be a null line K through a that is distinct from L and M (Figure 2.39).

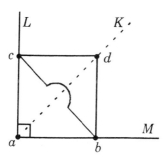

Figure 2.39

Take a point $d \neq a$ on K, and by taking parallels to L and M through d form the parallelogram $abcd$ as shown. Then the diagonals ad and bc are parallel. Since ad is null, this means (Corollary 2.3.4) that $ad \perp bc$, and so $abcd$ is a square in α_F. Theorem 2.6.5 then gives the desired result.

(2). If (1) fails, then $L \perp M$ implies $L \parallel M$. But then if $L \parallel M$ and a is a point on M,

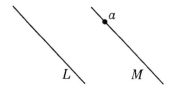

Figure 2.40

the altitude to L through a cannot intersect L, and must therefore be M itself, i.e. $L \perp M$. Hence "orthogonal" is synonymous with "parallel", and the characterisation of $\perp$ by

$$\begin{pmatrix} 0 & 1 \\ 1 & 0 \end{pmatrix}$$

is given as at the end of §2.5. $\qquad\qquad\qquad\qquad\qquad\qquad\qquad$ □

Our next concern is with the uniqueness of anisotropic planes. For this we need the notion of a *quadratic* field F, one in which, for each $k \in F$,

either $\sqrt{k}$ *or* $\sqrt{-k}$ *exists in F*

(the name "Euclidean" is often used for this condition, but that word is already doing enough work for us).

Theorem 2.6.8. *If F is a quadratic field, there is at most one anisotropic metric plane over F.*

Proof. Suppose $\alpha_F(-k)$ is anisotropic. Then by Corollary 2.6.4, $\sqrt{k}$ does not exist in F, so $\sqrt{-k}$ must exist. But then by Theorem 2.6.5, $\alpha_F(-k)$ is isomorphic to $\alpha(1)$. Thus all anisotropic planes over F (if there are any) are isomorphic. $\qquad\qquad\qquad\qquad\qquad$ □

Notice that the proof of Theorem 2.6.8 shows that if a quadratic field F gives rise to any anisotropic plane, then it must be $\alpha_F(1)$. But the latter is anisotropic just in case -1 has no square root in F (Corollary 2.6.4). In other words:

Corollary 2.6.9. *If F is quadratic, then there is a (necessarily unique) anisotropic plane over F if, and only if, $\sqrt{-1}$ does not exist in F.* $\qquad\qquad\qquad\qquad\qquad\qquad\qquad\qquad\qquad\qquad$ □

Since **R** is quadratic, it follows from the discussion of the last three sections that the Robb (singular), Lorentz (isotropic), and Euclidean (anisotropic) planes are all the metric planes that there are over the real number field. In the next section, abstract geometrical axioms will be given to characterise each of these three geometries.

As a partial converse to Theorem 2.6.8 there is

Theorem 2.6.10. *Suppose that $\sqrt{-1}$ does not exist in F. Then if there is at most one anisotropic plane over F, F is a quadratic field.*

Proof. If $\sqrt{-1}$ does not exist, then $\alpha_F(1)$ is anisotropic (Corollary 2.6.4), so in fact there is exactly one anisotropic plane over F, by the hypothesis. So if $\sqrt{k}$ does not exist, $\alpha_F(-k)$ is anisotropic, hence isomorphic to $\alpha_F(1)$. Theorem 2.6.5 then implies that $\sqrt{-k}$ must exist.   □

The hypothesis of the non-existence of $\sqrt{-1}$ cannot be dispensed with in Theorem 2.6.10, as may be seen when F is the field $\mathbf{Z}_5 = \{0, 1, 2, 3, 4\}$ of integers modulo 5. The non-zero elements with square roots are $1 = 1^2 = 4^2$ and $4 = 2^2 = 3^2$. Since $4 = -1$, $\sqrt{-1}$ does exist. However $\mathbf{Z}_5$ is not quadratic, since $-2 = 3$, and neither $\sqrt{2}$ nor $\sqrt{3}$ exist. On the other hand, the only possible anisotropic planes are $\alpha_{\mathbf{Z}_5}(-2)$ and $\alpha_{\mathbf{Z}_5}(-3)$, by Corollary 2.6.4, and these are isomorphic, by Theorem 2.6.2, since $-2 = -3 \cdot 2^2$ (because $2^2 = 4 = -1$). This example is in fact an instance of the following general situation.

Lemma 2.6.11. *If F is a finite field, then*
(1) *if F has characteristic 2, every non-zero element of F has a square root in F; and*
(2) *if the characteristic is not 2, there are elements of F without square roots. Moreover, if $\sqrt{j}$ and $\sqrt{l}$ do not exist in F, there is some $m \in F$ with $j = l \cdot m^2$.*

Proof. This is a standard piece of field theory, which is shown by using the properties of the squaring function $f(x) = x^2$ as a homomorphism from the multiplicative group $F^+ = F - \{0\}$ of non-zero elements of F to itself. The kernel of f is $H = \{1, -1\}$, and its image is the set

$$G = \{m^2 : m \in F^+\} = \{k : \sqrt{k} \text{ exists in } F\},$$

which is thereby a subgroup of F^+ isomorphic to F^+/H.

(1). If the characteristic of F is 2, then $1 = -1$, so $H = \{1\}$, implying that f, having trivial kernel, is one-to-one. But a one-to-one map from a finite set to itself is surjective (pigeonhole principle), so that $G = F^+$ as required.

(2). If $1+1 \neq 0$, H has 2 elements. Thus if F^+ has p elements, then F^+/H, and hence G, has $p/2$ elements. So, G has just two cosets in F^+, namely G itself, and

$$F^+ - G = \{k : \sqrt{k} \text{ does not exist}\}.$$

But this second coset is non-empty, since it has $p/2$ elements, and in this case $p \geq 2$. This shows that there is an element without a square root (indeed $p/2$ of them). Moreover, if $\sqrt{j}$ and $\sqrt{l}$ do not exist, then

$$jG = F^+ - G = lG,$$

so that $j \in lG$ as required.

Theorem 2.6.12. *If F is a finite field, then*
(1) *if F has characteristic 2, there are no anisotropic metric planes over F, and*
(2) *if $1 + 1 \neq 0$ in F, there is exactly one anisotropic plane over F.*

Proof. . From Lemma 2.6.11, by Corollary 2.6.4 and Theorem 2.6.2.

2.7 The Three Real Metric Planes

There are a number of basic facts of Euclidean geometry which seem obvious visually but which cannot be proven without a theory of *order*, i.e. of relative position of points. A notion of order is needed to define interiors and exteriors of circles, triangles etc., and in the absence of rather strong assumptions about order, such as Dedekind continuity, there can be no proof of such "facts" as that a line passing through the interior of a circle meets it in two different points, or that two circles with centres a, b and radius $|ab|$ actually have a point in common, as they appear to do when drawn (Figure 2.41). But the latter result is assumed in the very first proposition of Euclid's elements!

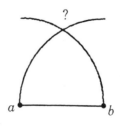

Figure 2.41

It was not until around 1880 that an explicit axiomatic analysis of order in the plane was developed. This was done initially by Pasch, using a three-placed relation of *linear betweenness* of points. This section will outline the way in which this relation gives rise to a set of axioms that uniquely characterises the real affine plane, and then go on to obtain characterisations of the Euclidean, Lorentz, and Robb planes.

Ordered Fields

An *ordered field* is a field F with a distinguished subset P that is closed under addition (+) and multiplication (·), and satisfies the *trichotomy law* that for each $x \in F$, exactly one of the conditions

$$x \in F, \quad -x \in F, \quad x = 0,$$

is true. P is to be thought of as a set of "positive" elements. Defining $x < y$ to mean that $(y - x) \in P$ then specifies $<$ as a *strict linear ordering* of F satisfying

$$0 < x, y \quad \text{implies} \quad 0 < x \cdot y,$$

$$x < y \quad \text{implies} \quad x + z < y + z.$$

(Conversely, any strict linear ordering of F satisfying these conditions is the ordering induced by taking $P = \{x : 0 < x\}$ as a set of positive elements.)

An ordered field F must be infinite, since it has $x < x + 1$, and so has

$$0 < 1 < \cdots < n < \cdots \cdots,$$

where the natural number n is identified with the F-element

$$n = \overbrace{1 + \cdots + 1}^{n \text{ times}}.$$

Then applying the field operations to the **n**'s, it is possible to construct a subfield

$$\{q : q \in Q\}$$

of F that is isomorphic to the rational number field **Q**.

The property that distinguishes **R** amongst ordered fields is that which ensures that there are no "gaps" in the number line:

DEDEKIND CONTINUITY AXIOM. *Let ordered field F be the union of two non-empty sets C and D such that $x < y$ for all $x \in C$ and $y \in D$. Then there is some $z \in F$ such that either*

(1) $z \in C$ and $x \leq z$, i.e. $x < z$ or $x = z$, for all $x \in C$; or
(2) $z \in D$ and $z \leq y$ for all $y \in D$
(more briefly: $x \leq z \leq y$ for all $x \in C$ and $y \in D$).

An ordered field satisfying this axiom is called *continuously ordered* (or *order-complete*). There is essentially only one such field, for any continously ordered field is isomorphic to **R**. In fact something even stronger can be shown: any continously ordered division ring is isomorphic to **R**, and hence is a field. The notion "continuously ordered" is defined for a division ring D just as for a field. If D is continously ordered, then it is *Archimedean*, which means that for any $x \in D$ there is some positive integer n with $x < $ **n**. This implies that the rational numbers are dense in D, i.e. if $x < y$ in D, then $x < \mathbf{q} < y$ for some $q \in \mathbf{Q}$. Then for any $x \in D$, the set of rationals $\{q : \mathbf{q} < x\}$ has a least upper bound r_x in **R**. The function $f(x) = r_x$ then gives the isomorphism between D and **R**. The proof of this does not require commutativity of multiplication in D.

It is noteworthy that the essential role of continuity in the argument just sketched is to ensure that f is surjective. Only the weaker Archimedean property is needed to ensure that f is a well-defined injective field homomorphism. Thus the argument shows that any Archimedean ordered division ring is isomorphic to a subfield of **R**, and hence is commutative. A proof of this along the present lines is presented in Artin [1957], while a direct algebraic derivation of the commutative law from the Archimedean property in ordered division rings appears in Hilbert [1971], §32.

Ordered Planes
If F is ordered by relation $<$, then a ternary relation $B(abc)$ can be introduced between triples of distinct points in the affine plane α_F. The expression $B(abc)$ is read "b lies between a and c", and is defined to mean that

$$\text{there exists } \lambda \in F \text{ with } 0 < \lambda < 1 \text{ and } b = (1 - \lambda)a + \lambda c \qquad (\dagger)$$

(recall that the general parametric form of a point on the line ac is $a + \lambda(c - a) = (1 - \lambda)a + \lambda c$).

The relation B satisfies the following properties, which are the classical betweenness axioms of Hilbert [1971].

B1. If $B(abc)$ then a, b, c are distinct collinear points with $B(cba)$.

B2. If $a \neq b$, there exists c with $B(abc)$.

B3. If a, b, c are distinct and collinear, then at most one of them
   lies between the other two.

B4. (PASCH'S LAW). If a line L passes between vertices a and b
   of a triangle, i.e. there is a point d on the line with $B(adb)$,
   and does not pass through the other vertex c, then L passes
   between a and c, or between b and c (Figure 2.42).

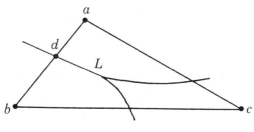

Figure 2.42

An *ordered* affine plane is one carrying a ternary relation B on its
set of points that satisfies B1 - B4. Such a plane always satisfies the
properties:

   if $a \neq c$, then there exists b with $B(abc)$; and

   if a, b, c are distinct and collinear, then exactly one of them lies
   between the other two

(cf. Hilbert[1971], Chapter 1,§4 for details.)

   Now if α is an ordered affine plane that is Pappian, and hence
coordinatised by a field F, then F can be ordered as follows. Recall
that the members of F are the points on a chosen line L of α, with
particular points o and e serving as the zero and unity of F (Figure
2.7). A set of *positive* elements $P \subseteq F$ is defined by putting

$$x \in P \quad \text{iff} \quad B(oxe) \text{ or } (x = e) \text{ or } B(oex),$$

i.e. $x \in P$ iff x is "on the same side of o as e". The ordering $<$
induced on F by P then defines a betweenness relation on α by the
condition (†) above, and this proves to be just the original relation
B we started with (for details of how this works, c.f. Ewald [1971],
§2.2, Appendix 3.III).

   The version of Dedekind continuity that is expressed in terms
of betweenness is:

B5. If the points of a line are divided into two non-empty disjoint subsets, neither of which posseses a point between two

points of the other, then there is a point of one set which
lies between every other point of that set and every point
of the other set.

An affine plane is *continuously ordered* if it has a betweenness relation satisfying B1 - B5. If α is continuously ordered and Desarguesian, and hence coordinatatised by some division ring D, then the
axiom B5 ensures that the relation $<$ induced on D by B1 - B4, in
the manner just described for a field-plane α_F, satisfies the Dedekind
Continuity axiom. Hence D is isomorphic to **R**. Consequently:

a continuously ordered Desarguesian affine plane is isomorphic
to the real affine plane α_R (and hence is Pappian).

This gives a complete "coordinate-free" characterisation of α_R, i.e. a
set of abstract geometrical properties which it and it alone satisfies.
The axiom system consisting of A1 - A3, B1 - B5, and the Desargues
Property is *categorical*, meaning that up to isomorphism there is
only one structure that satisfies all of these axioms simultaneously.

Now in the last section it was deduced that there are exactly
three non-null metric geometries based on α_R, and so we can now
give categorical axiomatisations of them as well.

Theorem. *Let* $M = (\alpha, \perp, B)$ *be a continuously ordered Desarguesian metric plane.*
(1). *If* M *is anisotropic (has no null lines), it is isomorphic to the
    Euclidean plane.*
(2). *If* M *is singular (has a singular line and a nonsingular line), it
    is isomorphic to the Robb plane.*
(3). *If* M *is isotropic (has a nonsingular null line), it is isomorphic
    to the Lorentz plane.*

3

Projective Transformations

This chapter develops an alternative method of coordinatising a metric affine plane, by embedding it into a *projective* plane, and using the orthogonality relation to define a matrix-representable transformation on the line at infinity. The construction will be central to our subsequent treatment of metric affine spaces of higher dimension, and its description now requires us to review a substantial body of additional ideas.

3.1 Projective Planes

On a couple of occasions in the previous chapter we alluded to the idea of parallel lines meeting at a "point at infinity". This can be made precise in the following way. Let $\alpha = (\mathcal{P}, \mathcal{L}, \mathcal{I})$ be an affine plane. The relation $L \parallel M$ of parallelism is an equivalence relation on $\mathcal{L}$. $\mathcal{P}$ is enlarged by adding one new point for each parallelism class, and then the set of all these new points is declared to be a new line L_∞ "at infinity". A set-theoretically convenient way to carry this out is to *define* L_∞ to be the set of equivalence classes (parallelism classes) of $\mathcal{L}$ under the relation $\parallel$. The new incidence structure can then be specified to be

$$\pi(\alpha) = (\mathcal{P} \cup L_\infty, \mathcal{L} \cup \{L_\infty\}, \mathcal{I}'),$$

where $a\mathcal{I}'L$ iff either

(1) $a\mathcal{I}L$ in α; or

(2) $a \in L_\infty$, and either $L = L_\infty$, or else $L \in a$ (i.e. the α-line L belongs to the parallelism class a).

It is then straightforward to show that $\pi(\alpha)$ satisfies the following properties.

P1. *Any two distinct points lie on exactly one line.*

P2. *Any two distinct lines pass through exactly one point.*

P3. *There exists a four-point, i.e. a set of four distinct points, no three of which are collinear.*

Any incidence structure satisfying P1, P2, P3, is called a *projective plane*. We have just seen that an affine plane α can be embedded in a projective plane $\pi(\alpha)$ - the *completion* of α. Conversely, if $\pi = (\mathcal{P}, \mathcal{L}, \mathcal{I})$ is a projective plane, any line L in π may be selected to act as the "line at infinity", and be deleted to give the incidence structure

$$\alpha(\pi - L) = (\mathcal{P} - \{a : a\mathcal{I}L\}, \mathcal{L} - \{L\}, \mathcal{I}),$$

which proves to be an affine plane. There is a natural isomorphism between π and the completion of $\alpha(\pi - L)$, which identifies L with the line at infinity of this completion by identifying point a on L with the parallelism class of all lines in α that meet at a in π.

Axioms P1 and P2 indicate that points and lines have equal "status" in projective planes, and leads to the *principle of duality*, stating that from any theorem provable from P1 - P3 another such theorem results immediately by interchanging the roles of points and lines. For instance, the dual of P3 states that there exist four lines, no three of which are concurrent. As the reader may care to verify, this can be derived from P1 - P3. Thus if $\pi = (\mathcal{P}, \mathcal{L}, \mathcal{I})$ is a projective plane, so too is the *dual plane*

$$\pi^{\partial} = (\mathcal{L}, \mathcal{P}, \mathcal{I}^{-1}),$$

where "point" L is incident with "line" a in π^{∂} if, and only if, $a\mathcal{I}L$ in π.

A property will be satisfied by π^{∂} precisely when its dual is satisfied in π. Thus if a property holds in all projective planes, then in particular it holds in all duals π^{∂}, and so its dual property holds in all projective planes π.

Desarguesian Planes

A pair of triangles (non-collinear triples of points) a, b, c, and a', b', c' are *centrally perspective* from a point p if corresponding vertices are collinear with p, i.e. if the lines aa', bb', and cc' all pass through p (Figure 3.1).

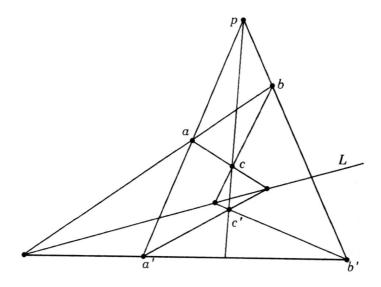

Figure 3.1

Dually, two triangles are *axially perspective* from a line L if corresponding sides are concurrent with L, i.e. if the points $ab \cap a'b'$, $ac \cap a'c'$, and $bc \cap b'c'$ all lie on L (strictly speaking, this dual statement should refer to *trilaterals* (non-concurrent triples of lines) associated with a, b, c and a', b', c'. Recall that the notation $L \cap M$ stands for the intersection point of lines L and M.)

A projective plane is *Desarguesian* if it has the

DESARGUES PROPERTY: *any two centrally perspective triangles are axially perspective.*

Notice that the dual of the Desargues property is its converse. In fact any Desarguesian plane satisfies this converse (cf. e.g. Garner [1981], p.73), so the principle of duality holds for the class of Desarguesian projective planes.

The Desarguesian property may be analysed more closely relative to a line L in a projective plane, as follows.

DESARGUES (L) PROPERTY: *if two triangles are centrally perspective from any point p not on L, and two pairs of corresponding sides meet on L, then the third pair of corresponding sides meet on L.*

LITTLE DESARGUES (L) PROPERTY: *if two triangles are centrally perspective from a point p on L, and two pairs of corresponding sides meet on L, then so does the third.*

The Desargues (L) property implies the Little Desargues (L) property (exercise: prove this by drawing a diagram, or see §5.2 of Stevenson [1972] for a more sophisticated proof involving transitivity of groups of collineations).

Recalling the Affine Desargues properties of §2.1 (Figures 2.4, 2.5), it is evident that an affine plane α is Desarguesian precisely when its projective completion $\pi(\alpha)$ has the Desarguesian (L_∞) property. Likewise, α has the Little Affine Desargues property precisely when $\pi(\alpha)$ has the Little Desargues (L_∞) property. Thus if π is any Desarguesian projective plane, and L is a line in π, the affine plane $\alpha(\pi - L)$ got by deleting L will be Desarguesian in the affine sense (since L may be regarded as the line at infinity in the completion of $\alpha(\pi - L)$).

Pappian Planes

PAPPUS (L, M, N) PROPERTY: *if a, b, c and a', b', c' are triples of collinear points on lines M and N, respectively, such that the points $ab' \cap a'b$ and $ac' \cap a'c$ lie on L, then $bc' \cap b'c$ lies on L as well* (Figure 3.2).

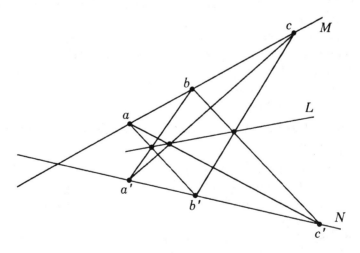

Figure 3.2

PAPPUS (L) PROPERTY: *the Pappus (L, M, N) property holds for all lines M and N distinct from L.*

A projective plane is *Pappian* if it has the Pappus (L) property for all lines L. Since the Pappus (L) property implies the Desargues (L) property (e.g. Stevenson [1972], §6.2), any Pappian projective plane is Desarguesian.

Recalling the Affine Pappus property of §2.1, we see that an affine plane α is Pappian precisely when its completion $\pi(\alpha)$ has the Pappus (L_∞) property. Thus if π is a Pappian projective plane, the planes of the form $\alpha(\pi - L)$ will all be Pappian affine planes.

Coordinates

Canonical examples of Pappian projective planes may be constructed out of the three-dimensional vector spaces F^3 over fields F. Let $\mathcal{P}$ be the set of all *lines* passing through the origin $\mathbf{o}$ in F^3, and $\mathcal{L}$ the set of *planes* of F^3 that pass through $\mathbf{o}$. If $p \in \mathcal{P}$ and $L \in \mathcal{L}$, define p to be *incident* with L if the line p lies in the plane L. This defines an incidence structure π_F which is readily seen to be a projective plane (any two lines (planes) through $\mathbf{o}$ are on a unique plane (line) through $\mathbf{o}$, etc.), and is moreover Pappian. Thus the bundle of affine lines and planes through the origin in F^3 becomes the collection of "points" and "lines" of π_F, which is known as the *projective plane over the point* $\mathbf{o}$, or the *plane at infinity of* F^3.

Now a line through the origin is a set of vectors of the form

$$p = \{\lambda b : \lambda \in F\},$$

where b is a particular (non-zero) direction vector for p. We may use b, or any of its scalar multiples λb as "coordinates" for the point p in π_F. Classically, particularly in relation to the real projective plane $\pi_{\mathbf{R}}$, it was customary to refer to points as being non-zero triples $b = (b_1, b_2, b_3)$ of "homogeneous" coordinates, and to regard any scalar multiple of b as determining the same point. Set-theoretically, we may consider the equivalence relation $\sim$ of *proportionality* on $F^3 - \{\mathbf{o}\}$, where

$$(x_1, x_2, x_3) \sim (y_1, y_2, y_3) \quad \text{iff} \quad \text{there exists } \lambda \in F \text{ with } x_i = \lambda \cdot y_i$$
$$\text{for } i = 1, 2, 3.$$

Observe that a *proportionality class*

$$[x_1, x_2, x_3] = \{(y_1, y_2, y_3) : (x_1, x_2, x_3) \sim (y_1, y_2, y_3)\}$$

(together with $\mathbf{o}$) is precisely a line through the origin in F^3. Thus we may take the points of π_F to be the set of such classes $[x_1, x_2, x_3]$.

A plane through the origin in F^3 is given by an equation

$$a_1 \cdot x_1 + a_2 \cdot x_2 + a_3 \cdot x_3 = 0, \tag{1}$$

with $a_1, a_2, a_3 \in F$ being not all zero. Replacing the coefficients of this equation by any proportional triple $(\lambda \cdot a_1, \lambda \cdot a_2, \lambda \cdot a_3)$ would give another equation for the same plane. Hence a line in π_F may be taken to be a proportionality class $< a_1, a_2, a_3 >$, with point $[x_1, x_2, x_3]$ being incident with point $< a_1, a_2, a_3 >$ precisely when equation (1) is satisfied. The use of the two different bracket notations $< \cdots >$ and $[\cdots]$ serves to distinguish points and lines in π_F. The fact that both are determined by non-zero triples of elements of F underscores the duality between points and lines in projective planes.

The association of the projective plane π_F with the three-dimensional geometry of F^3 provides a source of visual imagery that will guide much of what follows. For now it will be used to explain how π_F can be identified with (i.e. is isomorphic to) the projective completion $\pi(\alpha_F)$ of the affine field-plane α_F.

To begin with, α could be identified with a plane of F^3 that does not pass through the origin, say the horizontal plane $z = 1$. The correspondence $(x, y) \mapsto (x, y, 1)$ establishes this (Figure 3.3).

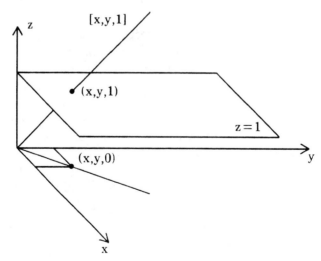

Figure 3.3

Alternatively, identify (x, y) with the projective point $[x, y, 1]$ in π_F (i.e. the line joining $(x, y, 1)$ to the origin). Thus the points of α_F

correspond to the lines through **o** that meet the plane $z = 1$, i.e. all lines through **o** except those, of the form $[x, y, 0]$, that lie in the x-y-plane. Any plane through **o** other than the x-y-plane will cut the plane $z = 1$ in a line, and so will correspond to a line in α_F. It then becomes clear that the x-y-plane itself corresponds to the line at infinity in the completion of α_F. For, given a line L through **o** in the x-y-plane, the bundle of all planes of F^3 containing L will cut $z = 1$ in a family of parallel lines. Moreover, any family of parallel lines in $z = 1$ arises from a line L of this type in the manner described, and the description establishes an exact correspondence between points of π_F of the form $[x, y, 0]$ and parallelism classes of lines in α_F, i.e. points at infinity in $\pi(\alpha_F)$.

This construction also provides a solution to the coordinatisation problem for any Pappian projective plane π. If L is a line in π, the structure $\alpha(\pi - L)$ is a Pappian affine plane, so is isomorphic to the plane α_F over some field F. Hence, as above, $\alpha(\pi - L)$ is isomorphic to the plane of all points $[x, y, z]$ with $z \neq 0$ in α_F. But then this isomorphism extends to one between π and π_F: each point p of L corresponds to a parallelism class of lines in $\alpha(\pi - L)$ (those lines having p as their point at infinity) and such a class of lines corresponds to a point of π_F of the form $[x, y, 0]$. In this way it is seen that any Pappian projective plane is isomorphic to a projective field-plane π_F. Furthermore, this *coordinatisation* can be set up in such a way that any prescribed line L becomes the line at infinity, i.e. the line $<0, 0, 1>$ of points of the form $[x, y, 0]$.

3.2 Projectivities and Involutions

If L is a line in a projective plane, the set $[L]$ of points on L is called a *range*, with *axis* L. If p is a point, the set $[p]$ of all lines passing through p is a *pencil*, with *centre* p.

If p is not on L, there is a natural bijection $f : [L] \rightarrow [p]$ obtained by joining each point of L to p, i.e. f maps point x on L to the unique line xp of the pencil $[p]$ that passes through x (Figure 3.4). f is called the *projection* of the range $[L]$ from p, and will be denoted $[L] \triangleright [p]$.

Dually there is the bijection $g : [p] \rightarrow [L]$ obtained by intersecting each line through p with L, i.e. $g(M) = M \cap L$ (Figure 3.5). g is the *section* of the pencil $[p]$ by L, and will be denoted $[p] \triangleleft [L]$. Clearly $[p] \triangleleft [L]$ is the inverse function to $[L] \triangleright [p]$.

A *projectivity* is a function obtained by composing a finite sequence of projections and sections. Such a sequence must consist *alternately* of projections and sections, since a projection can only

be composed with a section, and vice versa. A projectivity may map
a range or a pencil to a range or a pencil. It is always bijective, being
a composition of bijections, and its inverse is also a projectivity.

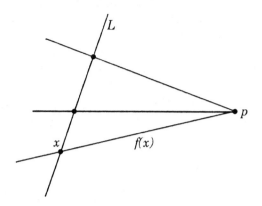

Figure 3.4

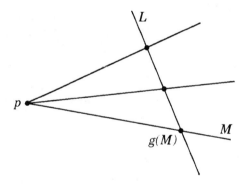

Figure 3.5

For the rest of this section we will concentrate on projectivities
that map a *range* to a *range*. Such a function must be made up
of projection-then-section pairs, and a pair of this type is called a
perspectivity. If point p is on neither of lines L and M, the *perspectivity from* $[L]$ *to* $[M]$ *with centre* p is the function $g : [L] \rightarrow [M]$
got by composing $[L] \triangleright [p]$ with $[p] \triangleleft [L]$, so that $g(x)$ is obtained by
letting the line through x and p meet M (Figure 3.6). This function
g will be denoted $[L] \overset{p}{\bowtie} [M]$. If $a, b, c \ldots$ are on L, and $a', b', c' \ldots$ are
on M, the notation

$$(a, b, c \ldots) \overset{p}{\bowtie} (a', b', c' \ldots)$$

indicates that the perspectivity with centre p maps a to a', b to b', c to c' etc. , i.e. that the lines aa', bb', cc', ... concur at p. Note that the inverse of $[L] \overset{p}{\bowtie} [M]$ is the perspectivity $[M] \overset{p}{\bowtie} [L]$.

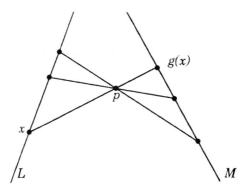

Figure 3.6

In any projective plane, a collinear triple (a, b, c) can be mapped to any other collinear triple (a', b', c') by a projectivity. Figure 3.7 indicates how to do this using two centres of perspectivity when the points are distinct and lie on different lines (the other cases are left as exercises).

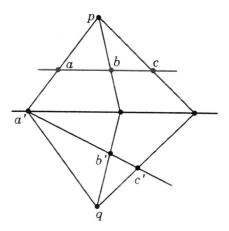

Figure 3.7

The question as to *how many* ways there are to construct a projectivity between given triples has a simple and significant answer in the case of Pappian planes:

Fundamental Theorem.

(1). *A projectivity is uniquely determined by three points and their images, i.e. if projectivities f and g assign the same values to three distinct points, then f = g.*

(2). *A projectivity that leaves three distinct points fixed is the identity function.*

These two statements are equivalent in any projective plane. That (1) implies (2) follows immediately from the observation that the identity function on any range of points is a projectivity fixing those points. Conversely, if f and g are projectivities agreeing on a, b and c, then $g^{-1} \circ f$ is a projectivity that fixes the three points, and so (2) implies that $g^{-1} \circ f$ is the identity function, making $f = g$.

The Fundamental Theorem is true in a projective plane if, and only if, that plane is Pappian. The derivation of the Pappus property from version (1) of the Theorem is indicated in Figure 3.8, which is generated by given triples a, b, c and a', b', c'.

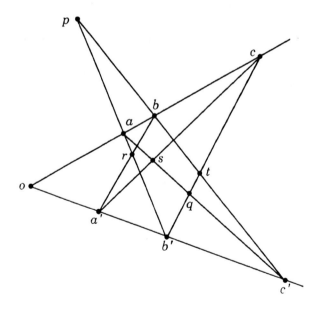

Figure 3.8

The object is to show that points r, s and t are collinear. Let f be the projectivity from $[pb']$ to $[ac']$ given by the sequence of perspectivities

$$(p, b', r, a) \overset{b}{\bowtie} (c', b', a', o) \overset{c}{\bowtie} (c', q, s, a).$$

But

$$(p, b', a) \overset{t}{\bowtie} (c', q, a),$$

i.e. f agrees on p, b', and a with the perspectivity with centre t, so by (1) f is this perspectivity, implying that, since $f(r) = s$, rs passes through t as desired.

To show that Pappian planes satisfy the Fundamental Theorem requires an extensive analysis of projectivities. The Desargues property is used in the proof to show that any projectivity can be expressed as a composite of at most three perspectivities, while the Pappus property is needed to show that a projectivity between two ranges that leaves their common point fixed must be a perspectivity. The details of the proof may be found, for example, in Garner [1981] Chapter 3, Hartshorne [1967] Chapter 5, Mihalek [1972] Chapter 7, or Stevenson [1972] Chapter 6.

Involutions

The type of projectivity that proves to be relevant to orthogonality relations is known as an *involution*. An involution is a non-identity projectivity $f : [L] \to [L]$ from a range to itself that is equal to its own inverse, which means that $f \circ f$ is the identity function on $[L]$. Such a mapping interchanges pairs of points on L, for if $f(a) = b$, then $f(b) = f(f(a)) = a$.

Now given any two pairs (a, a') and (b, b') of points on a line L, a projectivity can be constructed that interchanges the points of each pair. Figure 3.9 shows how to do this: take any point p not on L, and any line through a meeting pb in a point $q \neq p$. The composition of the sequence

$$[L] \overset{p}{\bowtie} [aq] \overset{b'}{\bowtie} [a'p] \overset{q}{\bowtie} [L] \tag{†}$$

then gives a projectivity with the desired effect.

In a Pappian plane any projectivity f that interchanges two distinct points, a and a', must be an involution. For if b is any other point on the line aa', there will be an involution interchanging (a, a') and $(b, f(b))$ (this works even if $f(b) = b$: the construction of Figure 3.9 is compatible with $b' = b$). But this involution agrees with f on the triple (a, a', b), and so by the Fundamental Theorem is identical to f. Notice that this argument shows that an involution is uniquely

determined by any two of its pairs (i.e. is the only projectivity interchanging those pairs), provided that at least one of the two pairs
consists of distinct elements.

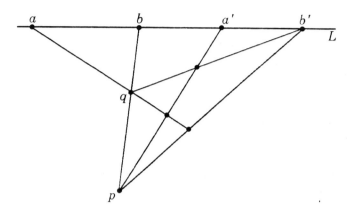

Figure 3.9

We finally come to the purpose of this whole discussion: the
association of an involution with an orthogonality relation. Let α
be a Pappian affine plane with a nonsingular orthogonality relation
$\perp$ satisfying the axioms O1, O2, O3 of §2.3. Let L_∞ be the line at
infinity in the projective completion $\pi(\alpha)$ of α. Define a function
$f_\perp : [L_\infty] \to [L_\infty]$ by a construction illustrated in Figure 3.10.

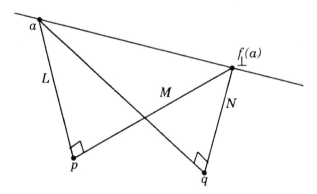

Figure 3.10

Here p is a fixed point in α. If a lies on L_∞, let L be the line in α
that passes through a and p in $\pi(\alpha)$. Then by the Altitude axiom
O2, there is a *unique* line M in α that passes through p and has

$L \perp M$. Let $f_\perp(a)$ be the point at infinity on M, i.e. the point where M meets L_∞ in $\pi(\alpha)$.

The definition of $f_\perp$ is independent of the choice of the point p, for if q is any other point in α, the line joining q to a in $\pi(\alpha)$ is parallel to L in α, hence its altitude N through q is parallel to M in α and meets M at $f_\perp(a)$ on L_∞ (Figure 3.10).

Assume that α has at least one non-null line under $\perp$, i.e. that $(\alpha, \perp)$ is not the degenerate Fano plane of §2.5. It follows that $f_\perp$ is not the identity function. It is also evident from the Symmetry axiom O1 that $f_\perp$ interchanges points on L_∞.

Involution Theorem. *$f_\perp$ is an involution on L_∞.*

Proof. Take two points a, b on L_∞, and let $a' = f_\perp(a)$, and $b' = f_\perp(b)$. Take pa here to be a non-null line, so that $a' \neq a$, and let g be the involution on L_∞ that interchanges the points of the pairs (a, a') and (b, b'). g is given by the sequence of perspectivities (†) above, as depicted in Figure 3.9. It will be shown that g is identical to $f_\perp$.

Take any point c on L_∞ and let $d = g(c)$. Generate the config-uration of Figure 3.11, an elaboration of Figure 3.9.

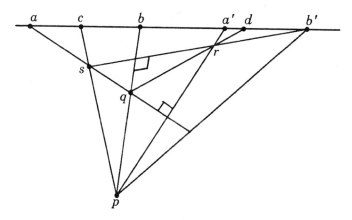

Figure 3.11

To show that $d = f_\perp(c)$, it is enough to show that $pc \perp pd$ in α. Consider the quadruple $pqrs$.

1. $pq \perp rs$. For, by construction $pq \perp pb'$, since pq is pb and $f_\perp(b) = b'$, and $pb' \parallel rs$, since they meet on L_∞, so $pq \perp rs$ by Theorem 2.3.1.

2. Similarly $pr \perp qs$, since $pr \perp pa$ and $pa \parallel qs$.

Thus by the Quadruple axiom O3, it follows that $ps \perp qr$. Now under the sequence (†) of perspectivities that make up g, c projects through p to s, which in turn goes to r through b', and finally r goes through q to $g(c)$. Since $g(c) = d$, the last step in this sequence entails that the line qr meets L_∞ at d. But then $qr \parallel pd$ in α. Since $ps \perp qr$, and $ps = pc$, this implies $pc \perp pd$.　　　　□

In the classical derivation of Euclidean plane geometry from real projective geometry, a certain involution is taken on the line at infinity in the real projective plane $\pi_{\mathbf{R}}$, and then two lines in $\alpha_{\mathbf{R}}$ are defined to be *perpendicular* if their ideal points (points at infinity) are interchanged by this involution (Coxeter [1949], Chapter 9). The axiomatic approach taken here in effect reverses that procedure. In this light the Quadruple axiom is seen to correspond to the following property of involutions, attributed to Pappus.

(‡)　*The three pairs of opposite sides of a complete quadrangle meet a line (not through a vertex) in three pairs of an involution*

(cf. e.g. Coxeter [1949], p.49).

A *complete quadrangle* is the six-lined figure formed obtained by taking a four-point a, b, c, d for the vertices and joining all pairs of these vertices (Figure 3.12). Opposite sides are those that meet at a point other than a vertex. The three points at which pairs of opposite sides meet are called the *diagonal* points of the quadrangle.

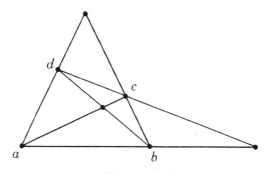

Figure 3.12

Suppose that the points a, b, c, d of Figure 3.12 lie in a metric plane $(\alpha, \perp)$, with $ab \perp cd$ and $ac \perp bd$. On the line L_∞ in the completion $\pi(\alpha)$ of α, there is an involution that interchanges the ideal points of the pairs (ab, cd) and (ac, bd) of opposite sides. The proof of our Involution Theorem consisted in showing that this involution is the mapping $f_\perp$ generated by the orthogonality relation $\perp$. But then the conclusion $ad \perp bc$ given by the Quadruple axiom requires that the ideal points of the third pair (ad, bc) of opposite sides also correspond under the involution.

The result ($\ddagger$) holds in any Pappian projective plane π_F coordinatised by a field of characteristic other than 2. The proof is based on the construction of Figure 3.13.

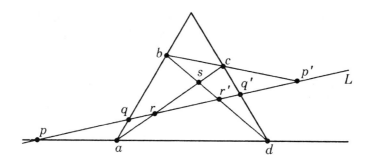

Figure 3.13

Here L is a line, not through a vertex, cutting the three pairs of opposite sides of the quadrangle $abcd$ in pairs (p, p'), (q, q'), and (r, r'). The assumption on the characteristic guarantees that the diagonal points of the quadrangle are not collinear. (For, if they were collinear, in the affine plane got by deleting their common line, $abcd$ would be a parallelogram with parallel diagonals, forcing $1 + 1 = 0$ in F - cf. the discussion of Fano planes in §2.1.) Thus L must fail to pass through at least one diagonal point, say the point s as drawn. It follows that $r \neq r'$.

Now let f be the projectivity given by the composition of

$$[L] \overset{a}{\bowtie} [bd] \overset{c}{\bowtie} [L].$$

f maps the sequence (p, q, r, r') to (q', p', r, r'). Let g be an involution on L interchanging the pairs (p', q') and (r, r'), and put $h = g \circ f$. Then h is a projectivity on L mapping (p, q, r, r') to (p', q', r', r). Since h interchanges a pair (r, r') of distinct points, the Fundamental Theorem for Pappian projective planes dictates that h must be involution (as noted above), and so it interchanges (p, p') and (q, q') as well.

Classically, the property ($\ddagger$) is used to derive the Euclidean theorem that the altitudes of a triangle concur (Coxeter [1949], p.118).

If an involution on a line L in a Pappian plane π is used to construct an orthogonality relation on the affine plane $\alpha(\pi - L)$ by the Euclidean definition, then to verify the Quadruple axiom it is precisely the involution property ($\ddagger$) that is needed, together with the fact that the involution is uniquely determined by any two of its pairs.

3.3 Matrix-Induced Projectivities

The Involution Theorem of the last section will now be used to coordinatise a nonsingular metric plane, by invoking the theory of matrix representations of involutions.

Let L_∞ be the line $< 0, 0, 1 >$ in a projective field-plane π_F. Each point on L_∞ has the form $[x, y, 0]$, which may be abbreviated to $[x, y]$. If

$$A = \begin{pmatrix} a & b \\ c & d \end{pmatrix}$$

is a 2×2 matrix over the field F, then A induces a function $f_A : [L] \to [L]$ by putting $f_A([x, y]) = [x', y']$, where

(1)
$$\begin{pmatrix} x' \\ y' \end{pmatrix} = A \begin{pmatrix} x \\ y \end{pmatrix},$$

i.e.

(2)
$$x' = a \cdot x + b \cdot y$$
$$y' = c \cdot x + d \cdot y.$$

f_A is well-defined on proportionality classes, for if (x, y) is replaced by $(\lambda \cdot x, \lambda \cdot y)$ in (1), then (x', y') gets replaced by $(\lambda \cdot x', \lambda \cdot y')$. The identity matrix

$$I = \begin{pmatrix} 1 & 0 \\ 0 & 1 \end{pmatrix}$$

induces the identity function on $[L]$, and the composite $f_A \circ f_B$ is induced by the matrix product AB. If A is invertible (when $a \cdot d \neq b \cdot c$) then f_A is a bijection of $[L]$, and conversely. The inverse function f_A^{-1} is induced by the matrix inverse A^{-1}.

Notice that, since $[x', y'] = [\lambda \cdot x', \lambda \cdot y']$, if $B = \lambda A$ then the functions f_A and f_B are identical. To prove the converse of this, a special role is played by the points $e_1 = [1, 0]$, $e_2 = [0, 1]$, and $u = [1, 1]$.

Lemma 3.3.1.
(1). If f_A fixes e_1, e_2, and u, then $A = kI$ for some $k \in F$, and so f_A is the identity function.
(2). If B is invertible, and f_A and f_B agree on e_1, e_2, and u, then $A = kB$ for some $k \in F$.

Proof.
(1). Since f_A fixes e_1, $[a, c] = [1, 0]$, so $c = 0$. Similarly $b = 0$. But then $[a, d] = f_A(u) = [1, 1]$, so for some k, $a = d = k$.
(2). $f_B^{-1} \circ f_A$ fixes e_1, e_2, u, so by (1), $B^{-1}A = kI$ for some k.

Corollary 3.3.2. If B is invertible and $f_A = f_B$, then $A = kB$ for some $k \in F$.

Theorem 3.3.3. If f_A is an involutory function, i.e. is a non-identity bijection equal to its own inverse, then $d = -a$.

Proof. If

$$B = \begin{pmatrix} d & -b \\ -c & a \end{pmatrix},$$

then f_B is the inverse of f_A, as B is a scalar multiple of matrix A^{-1}. Hence $f_B = f_A$, so $B = kA$ for some A.

If $a = d = 0$, then the result holds. Otherwise, since $d = k^2 \cdot d$ and $a = k^2 \cdot a$, we get $k^2 = 1$, so $k = \pm 1$. If $k = -1$, then $d = k \cdot a = -a$, and again the result holds. This leaves the case $k = 1$, in which $a = d$, $b = -b$, and $c = -c$. Then $b = c = 0$ is impossible, or else $A = aI$ and so f_A would be the identity function. But if, say, $b \neq 0$, then since $b = -b$, F must be of characteristic 2, so in this case $a = d = -d$.

Theorem 3.3.4. If (p, q, r) and (p', q', r') are triples of distinct points on $L_\infty = \langle 0, 0, 1 \rangle$, then there is a unique matrix-induced bijection of $[L_\infty]$ mapping the sequence (p, q, r) to (p', q', r').

Proof. Recall the representation of π_F as the family of lines through the origin **o** in the vector space F^3. Here p, q, r are distinct lines through **o** in the x-y-plane, so by choosing appropriate direction vectors for these lines we can obtain coordinates $[p_1, p_2]$ for p, $[q_1, q_2]$ for q, and $[r_1, r_2]$ for r such that

$$(p_1, p_2) + (q_1, q_2) = (r_1, r_2).$$

Then if

$$A = \begin{pmatrix} p_1 & q_1 \\ p_2 & q_2 \end{pmatrix},$$

it follows that A is invertible (as $[p_1, p_2] \neq [q_1, q_2]$), with $f_A(e_1) = p$, $f_A(e_2) = q$, and $f_A(u) = r$.

Similarly, there is an invertible matrix B such that f_B maps (e_1, e_2, u) to (p', q', r'). Then if $C = B(A^{-1})$, $f_C = f_B \circ f_A^{-1}$ maps (p, q, r) to (p', q', r') as desired.

Finally, if f_D agrees with f_C on $p, q,$ and r, then the matrices CA and DA induce mappings that agree on $e_1, e_2,$ and u. By Lemma 3.3.1(2) it then follows that $CA = kDA$ for some $k \in F$. Since A is invertible, this implies $C = kD$, and so $f_C = f_D$. $\qquad \square$

It has already been observed that the analogue of Theorem 3.3.4 for *projectivities* holds for any line in π_F. But the new point to be made now is that projectivities on $<0, 0, 1>$ turn out to be just the same thing as matrix-induced transformations f_A for invertible A. To see that the latter type of mapping is a projectivity, it is useful to introduce the "non-homogeneous" coordinate $z = x/y$ for the point $[x, y]$, provided that $y \neq 0$, i.e. that $[x, y] \neq [1, 0]$. The point $[1, 0]$ itself is assigned the coordinate $z = \infty$, so that the points of $<0, 0, 1>$ are identified with the members of $F \cup \{\infty\}$. Then the equations (2) for f_A become

(3) $$z' = \frac{a \cdot z + b}{c \cdot z + d}, \qquad (a, b, c, d \in F, \ a \cdot d \neq b \cdot c)$$

with the understanding that when $z = \infty$, z' is a/c if $c \neq 0$, and ∞ if $c = 0$. In the case that $c \neq 0$, this equation can be rewritten as

(‡) $$z' = \frac{a}{c} + \frac{b \cdot c - a \cdot d}{c} \cdot \frac{1}{c \cdot z + d}.$$

Functions given by equations of the type (3) are called *linear fractional transformations*, or *Möbius transformations*, and figure prominently in complex variable theory. (In the case $F = \mathbf{C}$, $\mathbf{C} \cup \{\infty\}$ is the *complex projective line*, which can be identified with the complex plane plus an additional point ∞, or, by stereographic projection of the plane, with a *Riemann sphere* having ∞ as the north pole. Linear fractional transformations correspond to conformal orientation-preserving transformations of the sphere.)

In the present context, the advantage of the notation (3) for f_A is that, in view of (‡), any linear fractional transformation can be obtained by composing transformations of the three simple types

(4) $$z' = z + b,$$

(5) $$z' = a \cdot z \qquad (a \neq 0),$$

(6) $$z' = \frac{1}{z},$$

and each of these types can be shown to be a projectivity in a geometrically natural and interesting way.

To see this, take a line J through the point $\infty = [1,0]$,with $J \neq L_\infty$, and consider the addition and multiplication of points on L_∞ in the affine plane $\alpha(\pi_F - J)$, by the method described in §2.1.

Type 4. Holding b fixed, the construction of $z' = z + b$ is indicated in Figure 3.14, which corresponds to Figure 2.8 with J as the line at infinity.

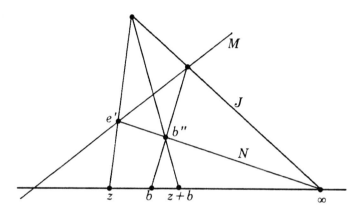

Figure 3.14

This shows that $z + b$ results from z by the projectivity

$$[L_\infty] \overset{e'}{\bowtie} [J] \overset{b''}{\bowtie} [L_\infty].$$

Type 5. Holding a fixed, $z' = a \cdot z$ results from z by the projectivity

$$[L_\infty] \overset{p}{\bowtie} [M] \overset{q}{\bowtie} [L_\infty]$$

of Figure 3.15, corresponding to Figure 2.9.

Type 6. When the projectivity

$$[L_\infty] \overset{e'}{\bowtie} [J] \overset{e}{\bowtie} [M] \overset{p}{\bowtie} [L_\infty]$$

in Figure 3.16 is applied to z, the result is a point z' with $z \cdot z' = e$, i.e. $z' = 1/z$ (as may be seen by putting $a = z$ and $z = z'$ in Figure 3.15).

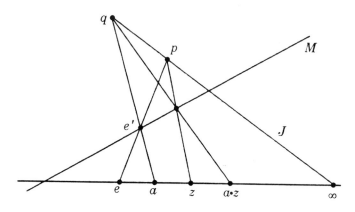

Figure 3.15

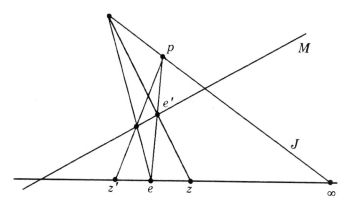

Figure 3.16

Theorem 3.3.5. *The projectivities on* $<0,0,1>$ *are precisely the matrix-induced transformations* f_A *for invertible A.*

Proof. Since any linear fractional transformation is a composition of mappings of the types (4), (5), and (6), and each of these types has been shown to be a projectivity, it follows that any matrix-induced transformation f_A on the line $<0,0,1>$ is a composition of projectivities, and hence is a projectivity. Conversely, if f is any projectivity on $<0,0,1!>$, then taking three distinct points p, q, r on $<0,0,1>$ (say $p = [1,0]$, $q = [0,1]$, and $r = [1,1]$), there is, by Theorem 3.3.4, a matrix A such that f_A maps (p, q, r) to $(f(p), f(q), f(r))$. But it has just been observed that f_A is a projectivity, and so by the Fundamental Theorem, f is identical to f_A, i.e. the projectivity f is matrix-induced.

Coordinatising a Metric Plane

Let $(\alpha_F, \perp)$ be a nonsingular metric affine field-plane (other than the degenerate Fano case). We now know that the involution $f_\perp$ induced on the line at infinity L_∞ of $\pi(\alpha_F)$ is of the form f_A for some matrix A. Moreover, by Theorem 3.3.3, A has the form

$$\begin{pmatrix} a & b \\ c & -a \end{pmatrix}.$$

Theorem 3.3.6. *The symmetric matrix*

$$G = \begin{pmatrix} c & -a \\ -a & b \end{pmatrix}$$

defines an inner product that characterises the orthogonality relation $\perp$.

Proof. Let $p = (x, y)$ be a point of α_F other than $\mathbf{o} = (0, 0)$. Under the identification of $\pi(\alpha_F)$ with π_F, L_∞ becomes the line $<0, 0, 1>$ and p becomes the point $[x, y, 1]$. The line joining p to $o = [0, 0, 1]$ in π_F is $< y, -x, 0 >$, and this line meets L_∞ at the point $[x, y, 0]$, which is being represented by $[x, y]$ (Figure 3.17).

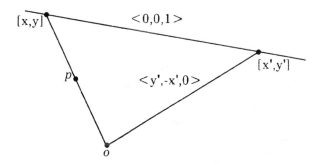

Figure 3.17

The altitudes to op all meet L_∞ at $f_\perp([x, y]) = [x', y']$, where

$$\begin{pmatrix} x' \\ y' \end{pmatrix} = \begin{pmatrix} a & b \\ c & -a \end{pmatrix} \begin{pmatrix} x \\ y \end{pmatrix}.$$

The altitude to op through o is the line $< y', -x', 0 >$ joining o to $[x', y']$. Thus for any other point $q = (z, w)$ in α_F, with $q = [z, w, 1]$ in π_F, we have

$$
\begin{aligned}
op \perp oq \quad &\text{iff} \quad q \text{ lies on } <y', -x', 0> \\
&\text{iff} \quad y' \cdot z - x' \cdot w = 0 \\
&\text{iff} \quad (y' \ -x') \begin{pmatrix} z \\ w \end{pmatrix} = 0.
\end{aligned}
$$

But since $x' = a \cdot x + b \cdot y$, and $y' = c \cdot x - a \cdot y$,

$$
(y' \ -x') = (x \ y) \begin{pmatrix} c & -a \\ -a & -b \end{pmatrix},
$$

and so

$$
\begin{aligned}
op \perp oq \quad &\text{iff} \quad (x \ y) \begin{pmatrix} c & -a \\ -a & -b \end{pmatrix} \begin{pmatrix} z \\ w \end{pmatrix} = 0 \\
&\text{iff} \quad pGq^T = 0
\end{aligned}
$$

$\square$

The construction in this Theorem provides another approach to the study of null lines, since these correspond to points on L_∞ that are left fixed by $f_\perp$. In the above notation (3) of linear fractional transformations, $z \in F \cup \{\infty\}$ is a fixed point of $f_\perp$ when

$$
(7) \qquad\qquad z = \frac{a \cdot z + b}{c \cdot z - a} \qquad (-a^2 \neq b \cdot c).
$$

But this is a quadratic equation in z, whose discriminant $4a^2 + 4b \cdot c$ is 0 when the characteristic of F is 2, and is non-zero otherwise. Thus when the characteristic is 2, there is exactly one fixed point z, hence one null line through each point in α_F (Artinian Fano plane). When the characteristic is not 2, then either there are no solutions to (7) (anisotropic case), or exactly two solutions, i.e. exactly two null lines through each point in α_F. In the latter case, if two intersecting null lines are taken as the coordinate axes, the points $[1,0]$ and $[0,1]$ will be fixed by $f_\perp$. But this requires that $[a, c] = [1, 0]$ and $[b, -a] = [0, 1]$, so $b = c = 0$. Then the inner product is given by matrix

$$
G = \begin{pmatrix} 0 & -a \\ -a & 0 \end{pmatrix},
$$

which induces the same orthogonality relation as does

$$\begin{pmatrix} 0 & 1 \\ 1 & 0 \end{pmatrix}.$$

This gives an alternative proof of the uniqueness of the isotropic geometry over a Pappian affine non-Fano plane.

In the anisotropic case, if a pair of orthogonal lines are used as coordinate axes, then the points $[1,0]$ and $[0,1]$ are *interchanged* by $f_\perp$, so $[a,c] = [0,1]$ and $[b,-a] = [1,0]$, giving $a = 0$ and

$$G = \begin{pmatrix} c & 0 \\ 0 & b \end{pmatrix}.$$

But then G induces the same orthogonality relation as does

$$\begin{pmatrix} c/b & 0 \\ 0 & 1 \end{pmatrix},$$

and we recover the form of inner product representation obtained in §2.6.

Matrices For Other Projectivities

The matrix representation of projectivities just developed is special to projectivities on $< 0,0,1 >$, since it depends on being able to present points in the form $[x,y]$. It is however possible to give a 2×2-matrix representation of projectivities between ranges on other lines, and between ranges of points and pencils of lines. For example, in the proof of Theorem 3.3.6 (Figure 3.17), let $g : [L_\infty] \to [o]$ assign to each point p on L_∞ the line joining o to $f_\perp(p)$. Then g is the composite of $f_\perp$ with the projection $[L_\infty] \triangleright [o]$, and so is a projectivity from the range of points on L_∞ to the pencil of lines centred on o. Each line through o has the form $<r,s,0>$, which may be abbreviated to $<r,s>$, and in these terms, $g([x,y]) = <y', -x'>$. But

$$\begin{pmatrix} y' \\ -x' \end{pmatrix} = \begin{pmatrix} c & -a \\ -a & -b \end{pmatrix} \begin{pmatrix} x \\ y \end{pmatrix},$$

showing that the projectivity g is induced by the matrix G which gave the inner product characterising $\perp$ (and providing a more geometrically significant account of the role of G).

This example is still special, because of the special form of lines in the pencil $[o]$. To represent projectivities in general by 2×2

matrices, it is necessary to assign coordinate *pairs* to points on lines and lines in pencils.

Let $a = [a_1, a_2, a_3]$ and $b = [b_1, b_2, b_3]$ be distinct points on a line L in π_F. Then for any point $x = [x_1, x_2, x_3]$ on L, there exist $r, s \in F$ with $x_i = r \cdot a_i + s \cdot b_i$, for $i = 1, 2, 3$. The pair (r, s) will be called a pair of *parameters* of x relative to the *base points* a and b on L.

The existence of r and s is clear geometrically, since (a_1, a_2, a_3), (b_1, b_2, b_3), and (x_1, x_2, x_3) are all non-zero vectors in the plane of F^3 corresponding to L, with the first two being linearly independent. Algebraically, consider the determinantal equation

(8)
$$\begin{vmatrix} x_1 & a_1 & b_1 \\ x_2 & a_2 & b_2 \\ x_3 & a_3 & b_3 \end{vmatrix} = 0.$$

This expands, by the first column, to a linear equation of the form

(9) $$p_1 \cdot x_1 + p_2 \cdot x_2 + p_3 \cdot x_3 = 0$$

over F in the "unknowns" x_i. Since (8), and hence (9), is satisfied by putting $x_i = a_i$, or by putting $x_i = b_i$, this must be the equation of the line L joining a and b in π_F. But from elementary linear algebra, if (8) holds then the first column is a linear combination of the other two.

Dually, since (9) holds with $x_i = a_i$, it follows that the *point* $p = [p_1, p_2, p_3]$ lies on the *line* $< a_1, a_2, a_3 >$. Similarly, p lies on the line $< b_1, b_2, b_3 >$. Then if $< x_1, x_2, x_3 >$ is any other line through point p, the x_i satisfy (9), hence (8), and so again there exist r and s with $x_i = r \cdot a_i + s \cdot b_i$ for $i = 1, 2, 3$. This shows that any line $< x_1, x_2, x_3 >$ in a pencil $[p]$ can be assigned a pair of parameters (r, s) relative to two chosen *base lines* $< a_1, a_2, a_3 >$ and $< b_1, b_2, b_3 >$ in that pencil.

Now let $[L]$ be a range with chosen base points a and b, and $[p]$ a pencil with chosen base lines M and N. Then any 2×2 matrix A induces a function $f_A : [L] \to [p]$ which maps the point q on L with parameters (r, s) relative to a and b to the line $f_A(q)$ through p with parameters (r', s') relative to M and N, where

(10) $$\begin{pmatrix} r' \\ s' \end{pmatrix} = A \begin{pmatrix} r \\ s \end{pmatrix}$$

(Figure 3.18).

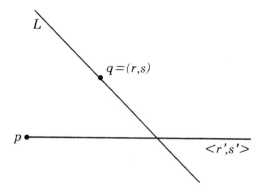

Figure 3.18

In the same manner, A determines mappings from $[p]$ to $[L]$, and between any two ranges, and any two pencils, relative to appropriately chosen base points and lines.

Consider the situation in which the chosen base points of $[L]$ are $a = [a_{11}, a_{12}, a_{13}]$ and $b = [a_{21}, a_{22}, a_{23}]$, and the chosen base lines of $[p]$ are $M = < m_{11}, m_{12}, m_{13} >$ and $N = < m_{21}, m_{22}, m_{23} >$. For $j, k = 1, 2$, put

$$x_{jk} = \sum_{i=1}^{3} m_{ji} \cdot a_{ki},$$

and let

(11)
$$A = \begin{pmatrix} -x_{21} & -x_{22} \\ x_{11} & x_{12} \end{pmatrix}.$$

Theorem 3.3.7. *Relative to a, b and M, N, f_A is the projection $[L] \triangleright [p]$ of $[L]$ from the point p.*

Proof. If point q on L has parameters (r, s), the parameters of the line $f_A(q)$ are (r', s'), where, by (10) and (11),

$$r' = -r \cdot x_{21} - s \cdot x_{22},$$
$$s' = r \cdot x_{11} + s \cdot x_{12}.$$

Hence
$$r' \cdot (r \cdot x_{11} + s \cdot x_{12}) = s' \cdot (-r \cdot x_{21} - s \cdot x_{22}),$$
and so

$$r' \cdot r \cdot x_{11} + r' \cdot s \cdot x_{12} + s' \cdot r \cdot x_{21} + s' \cdot s \cdot x_{22} = 0.$$

By definition of the x_{jk}'s, summation rules, and factorisation, this implies

$$\sum_{i=1}^{3}(r' \cdot m_{1i} + s' \cdot m_{2i}) \cdot (r \cdot a_{1i} + s \cdot a_{2i}) = 0,$$

which is precisely the condition for the point with parameters (r, s) relative to a and b to be incident with the line with parameters (r', s') relative to M and N, i.e. for q to lie on $f_A(q)$. Thus (cf. Figure 3.18) $f_A(q)$ is the line joining q to p, and so f_A is the projection of $[L]$ from p.

Corollary 3.3.8. *Any projectivity is the mapping f_A induced by some matrix A relative to suitably chosen base points/lines in its domain and range.*

Proof. The case of a projection $[L] \triangleright [p]$ is dealt with by the Theorem. The matrix A inducing such a projection must be invertible, since the projection is bijective. Then the inverse matrix of A induces the inverse function of the projection, which is the section $[p] \triangleleft [L]$ of the pencil $[p]$ by L. Since every section is the inverse of a projection, this takes care of the case of sections. But any projectivity is a composition of finitely many projections and sections, so the result holds in general, by induction, using the fact that $f_A \circ f_B = f_{AB}$. $\qquad\square$

Note that if the points $e_1 = [1, 0]$ and $e_2 = [0, 1]$ are taken as base points on $<0, 0, 1>$, then the point $[x, y]$ gets parameters (x, y) relative to e_1 and e_2. Hence the function f_A induced by an invertible A on $<0, 0, 1>$ *relative to e_1 and e_2* is just the mapping f_A defined at the beginning of this section, i.e. the mapping that was shown to be projective in Theorem 3.3.5.

Theorem 3.3.9. *The function f_A induced by an invertible matrix A between any two ranges/pencils, relative to chosen base points/lines, is always a projectivity.*

Proof. Take the case that f_A maps a range $[L]$ to a range $[M]$ relative to base points a, b on L and c, d on M. Let u and v be points not on L, M, or $L_\infty = <0, 0, 1>$, and construct the perspectivities $g : [L_\infty] \overset{u}{\bowtie} [L]$ and $h : [M] \overset{v}{\bowtie} [L_\infty]$ (Figure 3.19).

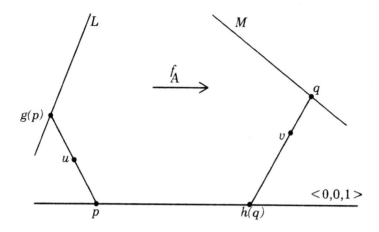

Figure 3.19

Let $j = h \circ f_A \circ g$. Now by Corollary 3.3.8, relative to the base points e_1, e_2 on L_∞, and a, b on L, the projectivity g is the function f_B for some matrix B. Likewise, relative to c, d and e_1, e_2, h is f_C for some matrix C. Thus

$$j = f_C \circ f_A \circ f_B = f_{CAB} : [L_\infty] \to [L_\infty]$$

relative to the base points e_1, e_2 on L_∞. By the note preceding this Theorem, this shows that j is just the function on $[L_\infty]$ induced by the matrix CAB, so by Theorem 3.3.5, j is a projectivity. Hence as

$$f_A = h^{-1} \circ j \circ g^{-1},$$

f_A is a projectivity too.

Similar proofs work in the other cases, when the domain of f_A is a pencil $[p]$ and/or the image of f_A is a pencil $[q]$. Then one takes g to be the projection $[L_\infty] \triangleright [p]$ and/or h to be the section $[q] \triangleleft [L_\infty]$.

3.4 Projective Collineations

Recall that a *collineation* of an incidence structure $(\mathcal{P}, \mathcal{L}, \mathcal{I})$ is a bijection $f : \mathcal{P} \to \mathcal{P}$ that preserves collinearity. Thus f maps a range $[L]$ with axis L onto another range of points, whose axis will be denoted $f(L)$. This definition extends f to act as a bijection $L \mapsto f(L)$ from $\mathcal{L}$ onto $\mathcal{L}$, such that point p lies on line L if, and only if, $f(p)$ lies on $f(L)$. The inverse of any collineation is a collineation, as is the composition of two collineations.

A *projective* collineation of a projective plane π is a collineation f with the property that for each line L of π,

the restriction $f|_L$ of f to the range $[L]$ is a projectivity.

Theorem 3.4.1. *A collineation of a projective plane is projective if, and only if, it induces a projectivity on at least one line.*

Proof. Suppose $f|_L$ is a projectivity for some line L. Take any other line M, and any point p not on L or M. Then $f|_M$ proves to be the projectivity

$$[f(L)]\overset{f(p)}{\bowtie}[f(M)] \circ f|_L \circ [M]\overset{p}{\bowtie}[L],$$

since

$$[f(M)]\overset{f(p)}{\bowtie}[f(L)] = f|_L \circ [M]\overset{p}{\bowtie}[L] \circ f|_M^{-1},$$

as the reader may verify by noting that f maps collinear points p, a, b to collinear points $f(p), f(a), f(b)$ as in Figure 3.20.

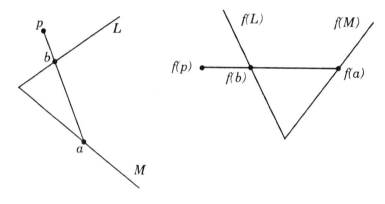

Figure 3.20

Theorem 3.4.2. *In a Pappian projective plane, the following hold.*
(1). *A projective collineation that fixes some four-point is the identity function.*
(2). *There is at most one projective collineation mapping a given four-point a, b, c, d onto a given four-point a', b', c', d'.*

Proof.
(1). Let a, b, c, d be a four-point (i.e. no three are collinear) that is fixed by projective collineation f. Let p be the intersection of $L = ab$ and $M = cd$. Now as $f(L) = f(a)f(b) = ab = L$, f maps any point q on L to a point $f(q)$ also on L. Similarly, if r is on M, so

is $f(r)$. Thus $f(p)$ lies on L and on M, i.e. $f(p) = p$. But then the projectivity $f|_L$ fixes the triple a, b, p on L, so by the Fundamental Theorem, $f|_L$ is the identity function on L. Similarly, f fixes all points on M. Finally then, if s is any point on neither L nor M, any two lines through s that are not through p will cut L and M in a four-point of fixed points under f (Figure 3.21).

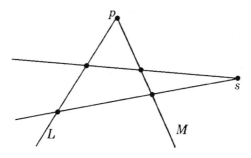

Figure 3.21

Repeating the above argument with this four-point will then show $f(s) = s$.

(2). If projective collineations f and g agree on a, b, c, d, then $g^{-1} \circ f$ is a projective collineation fixing a, b, c, d, so by (1), $g^{-1} \circ f$ is the identity function on the plane, making $f = g$.

Matrix-Induced Collineations

A 3×3 invertible matrix A over field F induces a bijection f_A of the points of π_F, by putting $f_A([x, y, z]) = [x', y', z']$, where

$$\begin{pmatrix} x' \\ y' \\ z' \end{pmatrix} = A \begin{pmatrix} x \\ y \\ z \end{pmatrix}.$$

If $A = kB$, for some $k \in F$, then $f_A = f_B$. By using the four-point $e_1 = [1, 0, 0]$, $e_2 = [0, 1, 0]$, $e_3 = [0, 0, 1]$, $u = [1, 1, 1]$, one can show, in a similar manner to Lemma 3.3.1, that

(1) if f_A fixes e_1, e_2, e_3, u, then $A = kI_3$, for some k, where I_3 is the 3×3 identity matrix; and

(2) if f_A and f_B agree on e_1, e_2, e_3, u, then $A = kB$ for some $k \in F$, and hence in particular

(3) if $f_A = f_B$, then $A = kB$.

Moreover, by adapting the argument of Theorem 3.3.4, using the present e_1, e_2, e_3, u, it can be shown that

(4) there is a unique matrix-induced bijection of π_F mapping a given four-point p, q, r, s onto a given four-point p', q', r', s'.

Theorem 3.4.3. *If A is invertible, the bijection f_A is a projective collineation.*

Proof. First, to see that f_A is a collineation, it is enough to show, for any three points $a_i = [x_i, y_i, z_i]$, that a_1, a_2, a_3 are collinear iff $f(a_1), f(a_2), f(a_3)$ are collinear. Let

$$B = \begin{pmatrix} x_1 & x_2 & x_3 \\ y_1 & y_2 & y_3 \\ z_1 & z_2 & z_3 \end{pmatrix}.$$

Then if the a_i are collinear, B has determinant $|B| = 0$ (cf. the discussion of equation (8) in the previous section). Hence matrix AB has zero determinant. But by definition of f_A, the columns of AB give homogeneous coordinates for the points $f(a_1), f(a_2), f(a_3)$, and so these points are collinear. Conversely, if the $f(a_i)$ are collinear, then $|AB| = 0$, and so, as $|A| \neq 0$, we must have $|B| = 0$, making the a_i collinear.

Next, to prove f_A is projective, it is enough, by Theorem 3.4.1, to show that f_A induces a projectivity on at least one line L. Take L to be $< 0, 0, 1 >$. If the i-j-th entry of A is a_{ij}, then f_A maps $[x, y, 0]$ on L to

$$[a_{11} \cdot x + a_{12} \cdot y, a_{21} \cdot x + a_{22} \cdot y, a_{31} \cdot x + a_{32} \cdot y],$$

and so

(†) $f_A([x, y, 0]) = [x(a_{11}, a_{21}, a_{31}) + y(a_{12}, a_{22}, a_{32})].$

In particular, the points $e_1 = [1, 0, 0]$ and $e_2 = [0, 1, 0]$ are mapped to the points $a = [a_{11}, a_{21}, a_{31}]$ and $b = [a_{12}, a_{22}, a_{32}]$. Then relative to the base points e_1, e_2 on L and a, b on $f_A(L)$, equation (†) shows that $f_A|_L$ maps the point with parameters (x, y) to the point with parameters (x, y), i.e. $f_A|_L$ is induced by the identity matrix, and hence is a projectivity by Theorem 3.3.9.

Corollary 3.4.4. *Any projective collineation of π_F is of the form f_A for some invertible matrix A.*

Proof. If f is a projective collineation, then by observation (4) preceding the Theorem, there is such a matrix-induced function f_A mapping the four-point e_1, e_2, e_3, u to the four-point

$$f(e_1), f(e_2), f(e_3), f(u).$$

By the Theorem, f_A is a projective collineation, and so by Theorem 3.4.2(2), $f = f_A$.

3.5 Correlations and Polarities

A *correlation* of a projective plane $\pi = (\mathcal{P}, \mathcal{L}, \mathcal{I})$ is an isomorphism f from π onto the dual plane $\pi^\partial = (\mathcal{L}, \mathcal{P}, \mathcal{I}^{-1})$. Thus f is a bijection from $\mathcal{P}$ onto $\mathcal{L}$ mapping collinear points in π to collinear points in π^∂, the latter being concurrent lines in π. Hence a correlation of π can be defined as a bijection $f : \mathcal{P} \to \mathcal{L}$ satisfying

p, q, r are collinear iff $f(p), f(q), f(r)$ are concurrent.

The image of a range $[L]$ under f will then be a pencil $[p]$ whose centre p will be denoted $f(L)$. This extends f to act as a bijection $L \mapsto f(L)$ from $\mathcal{L}$ onto $\mathcal{P}$ such that point p lies on line L if, and only if, line $f(p)$ passes through point $f(L)$. Thus the expression $g(f(p))$ is well defined when f and g are correlations, so any two such correlations can be composed, giving a function $g \circ f : \mathcal{P} \to \mathcal{P}$, which is a *collineation* ! Any correlation f has an inverse correlation $f^{-1} : \mathcal{P} \to \mathcal{L}$, with $f^{-1}(p)$ being that line L such that $f(L) = p$. Then $f^{-1} \circ f$ is the identity function on $\mathcal{P}$.

An alternative approach to this notion would be to define a correlation as a pair $f = (f_1, f_2)$ of bijections $f_1 : \mathcal{P} \to \mathcal{L}$ and $f_2 : \mathcal{L} \to \mathcal{P}$ such that in general p lies on L iff $f_1(p)$ passes through $f_2(L)$. Then the inverse of f would be (f_2^{-1}, f_1^{-1}), and the composition $g \circ f$ would be $(g_2 \circ f_1, g_1 \circ f_2)$.

A correlation is *projective* if its restriction $f|_L$ to any line is a projectivity from the range $[L]$ to the pencil $[f(L)]$. The composition of two projective correlations is a projective *collineation*, while the inverse of a projective correlation is another projective correlation. By dualising the appropriate parts of the proof of Theorem 3.4.1, the following is shown.

Theorem 3.5.1. *A correlation is projective if, and only if, it induces a projectivity on at least one line.*

Theorem 3.5.2. *A projective correlation is uniquely determined by its action on any four-point.*

Proof. If projective correlations f and g agree on four-point p, q, r, s, then $g^{-1} \circ f$ is a projective collineation fixing p, q, r, s, and so by Theorem 3.4.2(1), $g^{-1} \circ f$ is the identity function.   □

A 3×3 invertible matrix A over field F induces a bijection f_A between points and lines of π_F, by putting $f_A([x, y, z]) = <x', y', z'>$, where, as usual,

$$\begin{pmatrix} x' \\ y' \\ z' \end{pmatrix} = A \begin{pmatrix} x \\ y \\ z \end{pmatrix}.$$

By adapting the methods applied to collineations in §3.4, it is shown that

(1) f_A is a projective correlation;

(2) there is a unique matrix-induced correlation f_A mapping a given four-point p, q, r, s onto a given four-line P, Q, R, S ("four-line" meaning that no three of the lines are concurrent), and hence

(3) any projective correlation of π_F is the correlation f_A induced by some invertible matrix A.

Theorem 3.5.3. *The line-to-point mapping determined by the correlation f_A is induced by the matrix $(A^{-1})^T$.*

Proof. Write $<x>$ and $[x]$, respectively, for the line and point containing the triple $x = (x_1, x_2, x_3)$. Let $f_A([x]) = <x'>$, with $(x')^T = Ax^T$. For any line $<u>$, let $f_A(<u>) = [u']$. Since

$$[x] \text{ lies on } <u> \quad \text{iff} \quad [u'] \text{ lies on } <x'>,$$

the definition of incidence in π_F yields

$$ux^T = 0 \quad \text{iff} \quad u'(x')^T = 0,$$

and so

$$ux^T = 0 \quad \text{iff} \quad (u'A)x^T = u'(Ax^T) = 0. \tag{†}$$

Thus $<u>$ and $<u'A>$ are the same line (or else a point $[x]$ on one and not the other would falsify (†)). Hence u is proportional to $u'A$, so u' is proportional to uA^{-1}, and therefore $(u')^T$ is proportional to $(A^{-1})^T u^T$. Thus $(A^{-1})^T$ induces the mapping $<u> \mapsto [u']$.

Polarities

A projective correlation f is a *polarity* if it is involutory, i.e. is equal to its own inverse. This means that the function $f^{-1} : \mathcal{L} \to \mathcal{P}$ is just f (or that $f_2 = f_1^{-1}$, if f is thought of as a pair (f_1, f_2)). In other words, if $f(p) = L$, then $f(L) = p$. Thus the effect of a polarity is to interchange points and lines. The line $f(p)$ is called the *polar* of point p, and $f(L)$ is the *pole* of line L. From the incidence dualising property of f, it follows that if point p lies on the polar of point q, then q lies on the polar of p. Points related in this way are called *conjugate points*, and their polars are *conjugate lines*. A point that lies on its own polar is *self-conjugate*.

For an example of a polarity, take an invertible matrix A that is *symmetric*, i.e. $A^T = A$. Then $(A^{-1})^T = (A^T)^{-1} = A^{-1}$. Since A^{-1} induces the inverse $f_A^{-1} : \mathcal{L} \to \mathcal{P}$ to $f_A : \mathcal{P} \to \mathcal{L}$, Theorem 3.5.3 implies that f_A is a polarity. If A is instead *skew-symmetric*, i.e. $A^T = -A$, a similar argument would apply in general, were it not for the fact that in a field of characteristic $\neq 2$, a 3×3 skew-symmetric matrix

$$A = \begin{pmatrix} 0 & a & b \\ -a & 0 & c \\ -b & -c & 0 \end{pmatrix}$$

has determinant 0 and cannot be invertible. (In the characteristic 2 case, $A = -A$, and so symmetry and skew-symmetry come to the same thing.) These observations are needed to prove the following result.

Theorem 3.5.4. *If the correlation f_A is a polarity, then A is symmetric.*

Proof. Since the line-to-point mapping determined by f_A is induced by $(A^{-1})^T$, if f_A is a polarity, then the product matrix $(A^{-1})^T A$ induces the identity collineation, so $(A^{-1})^T A = kI_3$ for some k. Transposing this, $A^T A^{-1} = kI_3$, and so $A^T = kA$. Transposing again, $A = kA^T$. But then $A = k^2 A$, so $k = \pm 1$. If the characteristic is 2, then $k = 1$ and $A = A^T$ as desired. But if the characteristic is not 2, then $k \neq -1$, since, as noted above, A cannot be skew-symmetric. Hence again $k = 1$.

Historically, the notion of polarity arose from the study of *conics*. The conics first studied by the geometers of ancient Greece were the conic sctions: ellipse, hyperbola, and parabola. In the early nineteenth century, a projective definition was found by Steiner:

given a non-perspective projectivity $f : [p] \to [q]$ between pencils of lines on distinct centres, the associated conic C consists of all intersection points of corresponding lines under f, i.e. C is the set of points x of the form $f(L) \cap L$ (Figure 3.22).

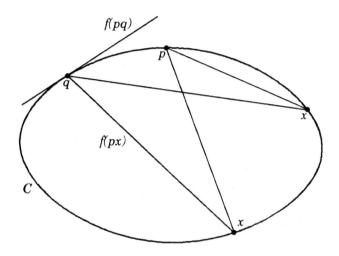

Figure 3.22

Now q is on the conic, since it lies on $f(pq)$. Moreover, no other point of C is on $f(pq)$ (as f is injective), so $f(pq)$ is a *tangent* to the conic at q. Inversely, $f^{-1}(pq)$ is tangent to the conic at p on C. It transpires that any two distinct points on C can be used to construct the conic in the manner described, and that through any point on the conic there is exactly one tangent to the conic.

Any line in the plane cuts a conic in at most two points, and so a conic is usually visualised in $\pi_{\mathbf{R}}$ as a simple closed curve, as in Figure 3.22, which appears in the real affine plane as an ellipse, parabola, or hyperbola, respectively, depending on whether it cuts the line at infinity in 0, 1, or 2 points.

A given conic determines a polarity as follows. If p is on the conic, the polar of p is the tangent to the conic at p. If p is not on the conic, take a quadrangle $abcd$ with vertices on the conic and p as one diagonal point (Figure 3.23). Then the polar of p is the line joining the other two diagonal points. This definition is independent of the choice of quadrangle: the polar is the locus of all points of intersection of the tangents to the conic at pairs of points that are collinear with p (cf. Veblen and Young [1910], Chapter V, for the theory of this construction).

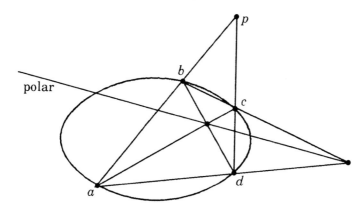

Figure 3.23

Conversely, one may take the approach (due to von Staudt) of defining a conic as the set of all self-conjugate (i.e. p lies on $f(p)$) points of a polarity f that has at least one such self-conjugate point. The tangents to this conic are then the self-conjugate lines under f.

Suppose that the polarity associated with a conic C is the transformation f_A induced by a symmetric matrix A. The points $[x]$ of C are those that lie on their own polars under f_A. But $[x]$ lies on $<Ax^T>$ if, and only if,

$$x A x^T = 0,$$

which expands to a homogeneous second degree equation

$$a_{11} \cdot x_1^2 + a_{22} \cdot x_2^2 + a_{33} \cdot x_3^2 + 2a_{12} \cdot x_1 \cdot x_2 + 2a_{13} \cdot x_1 \cdot x_3 + 2a_{23} \cdot x_2 \cdot x_3 = 0$$

for C of the type familiar from the analytic study of conic sections.

A polarity f induces an involution on any non-self-conjugate line L, with the fixed points of the involution being the points where L cuts the associated conic. The involution g maps point p on L to the intersection of $f(p)$ with L. Since the polar $f(p)$ of p must pass through the pole $f(L)$ of L (Figure 3.24), g is the composition of $f|_L$ with the section $[f(L)] \triangleleft [L]$ of the pencil $[f(L)]$ by the line L. Now $f|_L$ is a projectivity, since a polarity is by definition projective, and hence g is a projectivity. But since q is on $f(p)$ iff p is on $f(q)$, it follows that $q = g(p)$ iff $p = g(q)$, i.e. g is involutory, and corresponding pairs under g are conjugate under f.

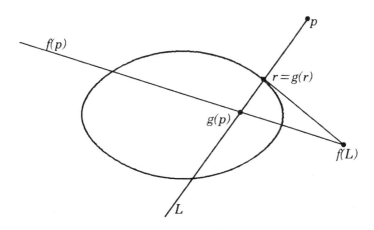

Figure 3.24

What does all of this have to do with orthogonality and space-time geometry? To explain that, let us return to the description of three-dimensional Minkowskian spacetime in Chapter 1. If L is a line through the origin, the lines through the origin that are orthogonal to L form a plane (cf. also Lemma 2.2.3). If L is timelike, the plane is a plane of simultaneity for an observer whose worldline is parallel to L. If L is lightlike, the plane is tangent to the lightcone and contains L. If L is spacelike, the plane is inertial (a copy of two-dimensional spacetime) and intersects the lightcone along two lightlike lines.

This assignment of planes to lines becomes an assignment of lines to points in the projective plane $\pi_{\mathbf{R}}$, and is in fact a polarity, hence is induced by a symmetric matrix G. This matrix defines an inner product, and therefore an orthogonality relation, on $\mathbf{R}^3$. But this is just the orthogonality relation of spacetime, and G is, up to a scalar multiple, the matrix giving the Minkowskian inner product. The conic of self-conjugate points in $\pi_{\mathbf{R}}$ defined by f_G is the set of lines through the origin in spacetime that lie in their associated planes. In other words, this conic is the *lightcone*. If α is any non-optical plane through the origin, then the involution induced by the polarity f_G on the non-self-conjugate "line" α in $\pi_{\mathbf{R}}$ is the same as the involution on the line at infinity of α given by the Minskowskian orthogonality relation in α.

All of this suggests a way to coordinatise an abstract orthogonality relation on a three-dimensional affine space F^3: use the orthogonality relation to *define* a polarity of π_F, and from this obtain a symmetric matrix, hence an inner product, as in Theorem 3.5.4. This procedure will be taken up in the next chapter, but first we need to consider the ramifications of a peculiar feature of the proof of 3.5.4, namely its dependence on the fact that an invertible 3×3 matrix cannot be skew-symmetric. At the next dimension up, this does not hold. To coordinatise four-dimensional spacetime we will work in projective *three*-space, and construct a projective interchange of points and *planes* that is induced by a 4×4 matrix. But such an interchange could be induced by an invertible skew-symmetric 4×4 matrix (the reader should verify that there are such things), producing a situation in which every point lies on its associated plane and is self-conjugate.

To take account of this, a different approach to Theorem 3.5.4 will be adopted, using the notion of a *self-polar triangle*. A triangle is self-polar under a given correlation f if each side is the f-image of the opposite vertex. Thus, as in Figure 3.23, the diagonal triangle of a quadrangle with vertices on a conic is self-polar under the polarity defined by that conic.

Theorem 3.5.5. *If a projective correlation f of a Pappian projective plane π has a self-polar triangle, then there is a coordinatisation of π relative to which f is induced by a diagonal (hence symmetric) matrix.*

Proof. Take one side L of the self-polar triangle as the line at infinity, and the other two sides as coordinate axes for the Pappian affine plane $\alpha(\pi - L)$. When α is coordinatised as α_F for some field F, π becomes π_F, and the self-polar triangle gets the coordinatisation of Figure 3.25.

As a projective correlation of π_F, f is induced by some matrix A. But since

$$f([1,0,0]) = < 1,0,0 >,$$

and similarly for the other vertices, A must have the form

$$\begin{pmatrix} a & 0 & 0 \\ 0 & b & 0 \\ 0 & 0 & c \end{pmatrix}$$

□

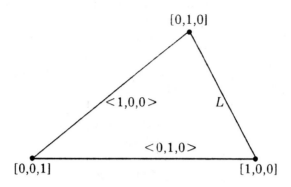

Figure 3.25

Now any polarity f of π_F must have a self-polar triangle (which, by 3.5.5, reiterates the result of 3.5.4 that f is induced by a symmetric matrix). To begin with, there must be a non-self-conjugate line P, for any two self-conjugate lines meet at a point p that is not self-conjugate. The reason for this is indicated in Figure 3.26,

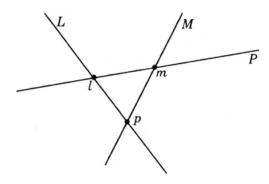

Figure 3.26

where l is the pole of L, and m is the pole of M. Since p lies on L, the polar P of p must pass through l. Similarly, P passes through m. But if P were self conjugate, all three points would have to coincide, producing distinct lines with the same pole, contrary to the bijectivity of f.

Now taking such a non-self-conjugate line P with pole p, the involution induced on $[P]$ by f has at most two fixed (self-conjugate) points, and so there exist distinct points q, r on P that are conjugate. Then the polar of q passes through r and p, while that of r passes through q and p. Hence pqr is a self-polar triangle.

At the next dimension up, the analogous concept is the self-polar *tetrahedron*, but it will be necessary to prove that a polarity has such a figure before a symmetric matrix representation can be obtained. More of that in Chapter 5.

4

Threefolds

In this chapter we give axioms for metric affine geometries of more than two dimensions, and then study the three-dimensional case in depth. The singular spaces are analysed first, and then the projective theory of Chapter 3 provides the basis for a coordinatisation of nonsingular threefolds.

4.1 Affine Spaces

The notion of an *incidence structure* will now be extended to that of a sequence $\Sigma = (\mathcal{P}, \mathcal{L}, \Theta, \mathcal{I})$, where

(1) $\mathcal{P}, \mathcal{L}$, and Θ are non-empty disjoint sets, whose members are called *points, lines,* and *planes*, respectively; and

(2) $\mathcal{I}$ is a binary relation of incidence, relating points to lines, lines to planes, and points to planes, and such that

$$p\mathcal{I}L \text{ and } L\mathcal{I}\alpha \text{ implies } p\mathcal{I}\alpha \qquad (\text{for } p \in \mathcal{P},\ L \in \mathcal{L},\ \alpha \in \Theta).$$

If line L is incident with plane α ($L\mathcal{I}\alpha$), then L is said to *lie in* α, while α *passes through* L, or *contains* L, etc. Lines L and M are *parallel*, $L \parallel M$, if either $L = M$, or else L and M lie in the same plane but do not intersect.

An incidence structure is an *affine space* if it satisfies the following axioms.

AS1. *Any two distinct points lie on exactly one line.*

AS2. *Any three non-collinear points lie in exactly one plane. Any plane contains three non-collinear points.*

AS3. *If point p is not on line L, then there is exactly one line M that passes through p and is parallel to L.*

AS4. *If L ∥ M and M ∥ N, then L ∥ N.*

Note that by the definition of incidence structure, if line L lies in plane α, then every point on L lies in α. Indeed L could be identified with $\{p : p\mathcal{I}L\}$, and α with $\{p : p\mathcal{I}\alpha\}$, so that lines and planes become sets of points, with set membership as the incidence relation. Furthermore, it would be possible (though more complicated) to dispense with planes as primitive entities, and to *define* the plane containing the non-collinear triple a, b, c as the set of all points on those lines that pass through one vertex of triangle abc and intersect the opposite side. (It is left as an exercise for the reader to determine what alternative axioms would then be needed to derive the above axioms from this definition.)

Axiom AS2 implies that if point p is not on line L, there is exactly one plane containing p and L: the one containing $p, q,$ and r, where q and r are distinct points on L. Likewise, if lines M and N intersect, there is exactly one plane containing them: the plane containing p, q and r, where p is the intersection point of M and N, and q and r are points on M and N distinct from p.

Now if points p and q lie in plane α, then all points on the line pq joining p to q lie in α (exercise: use AS2 and AS3). Then by AS1 - AS3, the points and lines incident with α satisfy the axioms of an affine plane. It is possible of course that Θ consists of a single element, in which case Σ is really just an affine plane. To guarantee that the structure Σ is *non-planar*, an additional axiom is needed:

AS5. *There exist four non-coplanar points.*

If points a, b, c, d are non-coplanar, then any three of them are non-collinear and determine a unique plane. This gives four planes, which are the *faces* of the *tetrahedron abcd*. The points a, b, c, d are the *vertices* of the tetrahedron. The *threefold* determined by this figure is the set of all points on those lines that pass through a vertex and intersect the opposite face. This threefold is also the smallest *subspace* of Σ containing a, b, c and d, where a *subspace* is a set S of points such that if $p, q \in S$, then all points on the line pq are in S.

In general, the smallest subspace containing a set T of points is said to be *generated* by T. Thus a line (onefold) determines a subspace generated by any two points on that line, while a plane (twofold) determines a subspace generated by any non-collinear triple lying in that plane, and a threefold is a subspace generated by a non-coplanar quadruple.

Any affine space satisfies the property

  AS4′: *if two planes in the same threefold intersect, their intersection is a line,*

and indeed AS4′ is equivalent to AS4 in any incidence structure satisfying AS1 - AS3 (cf. Sasaki [1952], Appendix, for the proof, which is lengthy).

To ensure that a non-planar affine space Σ is three-dimensional, we could make it a threefold by strengthening AS5 to require that there exists a non-coplanar quadruple that generates $\mathcal{P}$. Alternatively, AS1 - AS3, AS5 could be retained, and AS4 be replaced by

  AS4″: *any two intersecting planes intersect in a line.*

This ensures that any threefold contains all points of Σ, and gives a set of axioms that are essentially the "space" axioms of Hilbert [1971].

Coordinatisation

Let Σ be a non-planar affine space, i.e. Σ satisfies AS1 - AS5. Then Σ is *Desarguesian!*. In other words, the planar Desargues property becomes a theorem when more than two affine dimensions are assumed (for a proof of this celebrated result, cf., e.g., Ewald [1971], p.148). Σ can then be coordinatised as a vector space over a division ring. First of all, the points of any line L can be made into a division ring by carrying out the construction outlined in §2.1 within any plane of Σ that contains L. The result turns out to be the same division ring, up to isomorphism, no matter what line L and what plane containing L is chosen. Then the points of $\mathcal{P}$ are taken as vectors relative to some chosen point **o** as origin. The sum $p + q$ of vectors p and q is obtained by completing the parallelogram (Figure 4.1) in the plane determined by **o**,p, q, when these are not collinear. If **o**,p, q are collinear, then $p + q$ is given by addition in the division ring of their common line. Analogously, scalar multiples λp of a vector p are given by multiplication in the division ring of the line **o**p, under suitable identifications of the isomorphic division rings of lines.

This procedure makes $\mathcal{P}$ a (left) vector space over a division ring, whose lines and planes are just the original lines and planes of Σ. Details of this construction may be found in the paper by Bennett [1973], p.221.

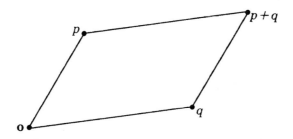

Figure 4.1

If Σ is further assumed to be Pappian, i.e. each plane of Σ is Pappian, then the coordinatising division ring will have commutative multiplication. Hence a Pappian affine space may be identified with the incidence structure of points (vectors), lines, and planes of a vector space over some field F. If this affine space is generated by a non-coplanar quadruple, the associated vector space will be three-dimensional, and so isomorphic to the canonical vector space F^3.

Since the division rings of lines in Σ are all isomorphic, if one is commutative then all will be. In view of the discussion in Chapter 2, this implies that for Σ to be Pappian it is enough that Σ have at least one pair of intersecting lines with the Pappus property.

4.2 Metric Affine Spaces

Let Σ be a non-planar affine space with a binary relation $\perp$ of orthogonality between its lines. A line L is *singular* in a plane α if L lies in α and is orthogonal to all lines in α. L is *nonsingular in α* if it lies in α but is not singular in α.

The structure $(\Sigma, \perp)$ is a *metric affine space* if it satisfies the following axioms.

OS1. *If $L \perp M$ then $M \perp L$.*

OS2. *If line L is nonsingular in plane α, then through each point p in α there passes exactly one line that lies in α and is orthogonal to L.*

OS3. *If $ab \perp cd$ and $ac \perp bd$, then $ad \perp bc$.*

OS4. *If lines L, M, N concur at point p, with $M \neq N$ and L orthogonal to both M and N, then L is orthogonal to every line through p in the plane containing M and N (Figure 4.2).*

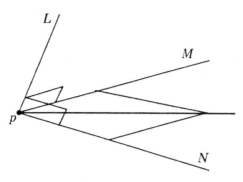

Figure 4.2

OS5. *If $L \perp M$ and $M \parallel N$, then $L \perp N$.*

A line L in Σ is *singular* if it is orthogonal to all lines in Σ, and *nonsingular* otherwise. $(\Sigma, \perp)$ is a *null space* if all lines in Σ are orthogonal to each other, i.e. all lines are singular lines. Σ is a *singular space* under $\perp$ if it contains at least one singular line, but is not a null space, i.e. has at least one nonsingular line as well. If all lines are nonsingular, then $(\Sigma, \perp)$ is a *nonsingular space*.

The incidence structure associated with any metric vector space $(V, \bullet)$ over a field is a metric affine space. Verification of axioms OS1 - OS3 has already been discussed in §2.2. For OS4, observe that if L, M, N have direction vectors l, m, n, respectively, then any line in the plane of M and N has direction vectors of the form $\lambda m + \mu n$. But $l \bullet (\lambda m + \mu n) = 0$ if $l \bullet m = l \bullet n = 0$.

Verification of OS5 is straightforward, since parallel lines have the same direction vectors. The point here is that our interpretation of the relation $L \perp M$ as meaning that L and M have orthogonal directions automatically makes $\perp$ closed under parallelism, in the sense of OS5. Some geometers might prefer to restrict $\perp$ to be a relation that holds only between coplanar lines, and so would define a singular line as one orthogonal to all lines with which it is coplanar etc. Then if $L \perp M$ in our present sense, M will be parallel to to some line M' with L orthogonal to M' in the more restricted sense. In order to develop the theory on the basis of this alternative interpretation, OS5 would have to be replaced by a more complex hypothesis, such as the following.

If $L \perp M$ and $L \parallel L'$, then $L' \perp N$ whenever $M \parallel N$ and N is coplanar with L'.

The main role of axiom OS5 in the present theory is to establish that the relation $\perp$ is determined by its restriction to the pencil $[p]$ of all lines in Σ that pass through any given point p:

Theorem 4.2.1. *The parallelism relation induces an orthogonality-preserving bijection between any two pencils $[p]$ and $[q]$.*

Proof. Let L be any line through p. If L is the line pq joining p to q, put $L' = L$. Otherwise, when q is not on L, let L' be the unique line through q that is parallel to L, as given by axiom AS3. If $L \perp M$ in $[p]$, then since $M \parallel M'$, OS5 gives $L \perp M'$. But $L' \parallel L$, so OS5 again gives $L' \perp M'$. Similarly, $L' \perp M'$ implies $L \perp M$. $\square$

Repeated use of the construction of this Theorem will be made to transfer the orthogonality relation from point to point, and to deduce such facts as that any line parallel to a singular line is singular, that parallel planes have isomorphic orthogonality structures, and so on. One particularly useful case of the construction gives:

Theorem 4.2.2. *If L is orthogonal to all lines through a point p on L, then L is singular.* $\square$

If an affine space Σ is coordinatised by a field of characteristic 2, then the relation of parallelism satisfies OS1 - OS5, and gives a non-null (indeed nonsingular) space in which all lines are null, and all planes are degenerate Fano planes (cf. §2.5). Also, over such fields the matrices

$$\begin{pmatrix} 1 & 0 & 0 \\ 0 & 1 & 0 \\ 0 & 0 & 1 \end{pmatrix} \qquad \begin{pmatrix} 1 & 0 & 0 \\ 0 & 0 & 1 \\ 0 & 1 & 0 \end{pmatrix}$$

define nonsingular metric threefolds with degenerate, isotropic, and singular planes. The reader is invited to explore these strange examples for himself: from now on we will rule them out, and assume that Σ satisfies Fano's axiom, i.e. the diagonals of any parallelogram intersect.

Observe that by OS1 - OS3, the restriction of $\perp$ to the lines of any plane α in Σ makes α into a metric affine plane. If α is nonsingular, it will have intersecting orthogonal lines, and so, as Σ is Desarguesian, α will be Pappian (Corollary 2.3.7). This implies that the division ring of points on some line is a field, which is enough, as noted at the end of the last section, to guarantee that Σ is coordinatised as a vector space over a field.

The upshot of this discussion is that although neither the De-sargues nor the Pappus properties are included in the axioms for a metric affine space, they are always present:

Theorem 4.2.3. *Any metric affine space with at least one nonsingular plane is Pappian.*

Theorem 4.2.4. *A nonsingular metric affine space has nonsingular planes through each point.*

Proof. Suppose that all planes in $(\Sigma, \perp)$ through p are either singular or null. Then there is a null line L through p in Σ. Let M be any line through p. If $M = L$, then $L \perp M$ as L is null. If $M \neq L$, then L and M generate a plane α containing p. Either α is a null plane, so immediately $L \perp M$, or else α is a singular plane, so L is a null line in a singular plane, hence is singular in that plane, and again $L \perp M$. Thus L is orthogonal to all lines through p, so by 4.2.2 is singular in Σ, contradicting the hypothesis of nonsingularity of the space.

4.3 Singular Threefolds

On $\mathbf{R}^3$ there are essentially three singular orthogonality relations.

1. The matrix

$$\begin{pmatrix} 1 & 0 & 0 \\ 0 & 0 & 0 \\ 0 & 0 & 0 \end{pmatrix}$$

induces the inner product on $\mathbf{R}^3$ assigning to points $a = (x_1, y_1, z_1)$ and $b = (x_2, y_2, z_2)$ the number

$$a \bullet b = x_1 x_2,$$

so that $oa \perp ob$ if, and only if, $x_1 = 0$ or $x_2 = 0$, i.e. if and only if at least one of a and b lies in the y-z-plane. The "light-cone" ($a \bullet a = 0$) is thus the y-z-plane, a *null* plane, all lines of which are singular (Figure 4.3). All other lines in $\mathbf{R}^3$ are orthogonal only to the lines in this null plane, so all other planes through the origin are singular.

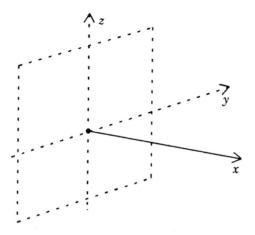

Figure 4.3

In terms of the criteria for nullity of a plane of Theorem 2.4.3, observe that the y-z-plane has intersecting lines that are singular in that plane.

2. The matrix

$$\begin{pmatrix} 1 & 0 & 0 \\ 0 & -1 & 0 \\ 0 & 0 & 0 \end{pmatrix}$$

induces the inner product

$$a \bullet b = x_1 x_2 - y_1 y_2,$$

so that $a \bullet a = x_1^2 - y_1^2$, and

$$a \bullet a = 0 \quad \text{iff} \quad y_1 = \pm x_1.$$

Thus the lightcone consists of the two planes $y = x$ and $y = -x$ (Figure 4.4), each of which is null. The z-axis is singular in $\mathbf{R}^3$, and is the only singular line through the origin **o**. Thus the two null planes fulfill the second criterion of Theorem 2.4.3, in having a singular line intersected by null nonsingular lines. Any plane containing the z-axis, other than the two null planes, is a singular plane with exactly one singular line through each point. Any plane through **o** that does not contain the z-axis is isotropic, and meets the light cone along two nonsingular null-lines.

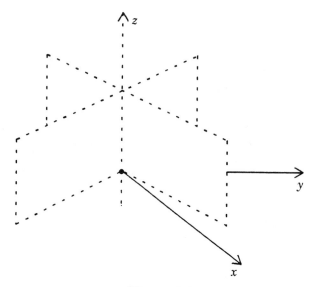

Figure 4.4

3. The matrix

$$\begin{pmatrix} 1 & 0 & 0 \\ 0 & 1 & 0 \\ 0 & 0 & 0 \end{pmatrix}$$

induces the inner product

$$a \bullet b = x_1 x_2 + y_1 y_2,$$

giving the geometry of the Robb threefold of Figure 1.15, in which the z-axis is singular, and is the only null line through the origin. Any plane containing the z-axis is singular, while any other plane through the origin is anisotropic.

Now consider an abstract singular threefold $(\Sigma, \perp)$. Σ is an affine space generated by four non-coplanar points, and is a non-null space that has a singular line. The distinctively three-dimensional property of Σ that will be used below is that any two intersecting planes meet in a line. Spaces of this type fall into three classes.

Type 1. There exist intersecting lines that are both singular in Σ.

Type 2. There exists a singular line that is intersected by a non-singular null-line.

Type 3. There is a singular line that is intersected only by non-null lines.

Analysis of Type 1 Geometries

Let M and N be intersecting singular lines in a singular threefold $(\Sigma, \perp)$, and let α be the plane of Σ in which M and N lie.

Theorem 4.3.1.

(1). *Any line in α is singular in Σ.*

(2). *Any null line intersecting α must lie in α.*

(3). *Any plane intersecting α is a singular plane whose null lines are all parallel to the line of intersection with α.*

Proof.

(1). Let M and N meet at o in α. If L is any line in Σ through o, then L is orthogonal to M and to N, since M and N are singular, and so by axiom OS4, L is orthogonal to all lines through o in α.

This shows that if K passes through o in α, then K is orthogonal to any line L through o, so K is singular (4.2.2). But any other line K' in α is parallel to a line K through o in α, and hence, using OS5, is singular as well.

(2). Let L be a null line meeting α. Suppose that L does not lie in α. Then L and α meet at a point o. Take any line K through o with $K \neq L$. L and K generate a plane β that is distinct from α, and so meets α in a line J (Figure 4.5).

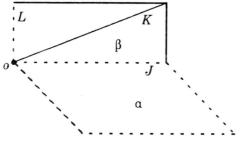

Figure 4.5

But J is singular, by part (1), and so β has a singular line (J) intersecting a null line (L). Theorem 2.4.3 then implies that β is a null plane, and so in particular K is a null line.

This establishes that all lines through o are null, so that any two lines through o lie in a null plane and are orthogonal to each other. But this makes $(\Sigma, \perp)$ a null space, contrary to hypothesis.

It follows that L must lie in α.

(3). Let β be a plane intersecting α in a line L. Then L is singular in β, by part (1). Any other line in β that intersects L must be non-null, by part (2). Thus β is a singular plane as described (cf. §2.4). □

This Theorem gives a complete description of the orthogonality relation $\perp$ in Σ, and shows it to be just as for the first of the three geometries on $\mathbf{R}^3$ listed above: the lines in α (and their parallels) are singular, and any other line that meets α is nonsingular and orthogonal only to the lines in α (and their parallels). Thus if α is used as the y-z-axis in a coordinatisation of Σ as a vector space F^3 over a field, the inner product induced by the matrix

$$\begin{pmatrix} 1 & 0 & 0 \\ 0 & 0 & 0 \\ 0 & 0 & 0 \end{pmatrix}$$

will characterise $\perp$.

This demonstrates that there is essentially only one Type 1 singular metric structure over any Pappian affine threefold.

Analysis of Types 2 and 3

Both of these cases will make use of the following construction. Let L be a singular line that meets a plane α in a point o. Suppose that α is a nonsingular plane, i.e. that no line of α is orthogonal to all lines in α. Then α, and indeed Σ, are Pappian (Theorem 4.2.3), and so by the work of Chapter 2, α can be coordinatised as a field-plane α_F, with o as the origin, and with a matrix of the form

$$G = \begin{pmatrix} d & 0 \\ 0 & e \end{pmatrix}$$

characterising the relation $\perp$ in α. This coordinatisation then lifts to Σ, by taking L as the z-axis and α as the x-y-plane, changing the coordinates (x, y) of a point in α to $(x, y, 0)$.

Now if $a = (x, y, z)$ is any point in Σ not on L, let $a' = (x, y, 0)$ be the "projection" of a onto α.

Lemma 4.3.2. $oa \perp ob$ if, and only if, $oa' \perp ob'$.

Proof. By construction, the plane β containing oa and L (the z-axis) meets α (the x-y-plane) along oa'. Likewise, the plane γ containing ob and L meets α along ob' (Figure 4.6).

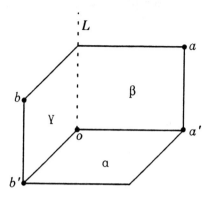

Figure 4.6

Suppose $oa \perp ob$. Then since $oa \perp L$ (L being singular), oa is orthogonal to all lines in γ, by OS4, and hence $oa \perp ob'$. But then $ob' \perp oa$ and $ob' \perp L$, so by OS4 again, $ob' \perp oa'$. The converse is similar.   □

Now let • be the inner product induced by the matrix

$$H = \begin{pmatrix} d & 0 & 0 \\ 0 & e & 0 \\ 0 & 0 & 0 \end{pmatrix}.$$

Then $a \bullet b = a' \bullet b'$ in general. But in view of the role of G in α_F, it follows that

$$oa' \perp ob' \quad \text{iff} \quad a' \bullet b' = 0.$$

By Lemma 4.3.2 then,

$$oa \perp ob \quad \text{iff} \quad a \bullet b = 0.$$

Since H makes the z-axis singular, this proves that H completely characterises $\perp$ in Σ.

Type 2
Let $(\Sigma, \perp)$ be a metric affine threefold having a singular line L that is intersected (at o) by a null line M that is itself not singular. Then there is a line N through o that is not orthogonal to M, and in the plane α containing M and N, M is a nonsingular null line. By §2.5,

α must then be an isotropic plane, whose orthogonality relation can be coordinatised by the matrix

$$G = \begin{pmatrix} 1 & 0 \\ 0 & -1 \end{pmatrix}.$$

Hence by the above construction, there is a coordinatisation of Σ in which $\perp$ is characterised by the matrix

$$H = \begin{pmatrix} 1 & 0 & 0 \\ 0 & -1 & 0 \\ 0 & 0 & 0 \end{pmatrix}.$$

This shows that over any coordinatising field, there is essentially only one Type 2 singular threefold, the one whose description matches that of the second of the list of cases over **R**.

Type 3

If the singular line L is not intersected by any null line, then any plane α intersecting L will be anisotropic, and have a coordinatising matrix of the form

$$\begin{pmatrix} d & 0 & 0 \\ 0 & 1 & 0 \\ 0 & 0 & 0 \end{pmatrix}.$$

If the coordinatising field is *quadratic* (i.e. each element or its negative has a square root), then it can be arranged that $d = 1$ (Theorem 2.6.8), and so over such a field there is essentially only one Type 3 singular threefold, one whose description matches that of the Robb threefold over **R**.

Notice that this analysis has shown that Type 2 and 3 singular threefolds always have nonsingular planes, and so are Pappian.

Criteria for Singularity of a Threefold

Theorem 4.3.3. *If line L is singular in a plane α in Σ, and L is orthogonal to some line that does not lie in α, then L is singular in Σ.*

Proof. Suppose $L \perp M$, with M not lying in α. We can suppose that L and M intersect, at o say. Let N be any line through o other than M. Since M is not in α, the plane containing M and N intersects α along a line K (Figure 4.7).

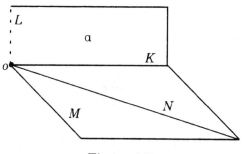

Figure 4.7

Then $L \perp K$, since L is singular in α, and $L \perp M$, so by OS4 L is orthogonal to all lines through o in the plane of M and K. Hence $L \perp N$.

This shows that L is orthogonal to all lines through o, and so is singular in Σ by Corollary 4.2.2.

Corollary 4.3.4. *If two planes intersect in a line L that is singular in each plane, then L is singular in Σ.*

Theorem 4.3.5. *A non-null metric affine threefold is singular if it satisfies either of the following conditions.*
(1). There exist two intersecting null planes.
(2). There exists a null plane intersected by a singular plane.

Proof.
(1). If null planes α and β intersect in line L, then L is singular in α and in β, so is singular in Σ by 4.3.4.

(2). If null plane α meets singular plane β in line L, then L is a null line in the singular plane β, and so by §2.4, L is singular in β. Since L is singular in α, 4.3.4 applies again.

4.4 Nonsingular Threefolds

Over $\mathbf{R}^3$ there are two nonsingular metric geometries: Euclidean three-space, and three-dimensional Minkowskian spacetime. This section will derive categorical axiomatisations of each.

Fix a metric affine threefold $(\Sigma, \perp)$ that is nonsingular. The affine space Σ may be coordinatised as the vector space F^3 over some field, and, since we are assuming that Σ satisfies Fano's axiom, the characteristic of F is not 2.

Theorem 4.4.1. $(\Sigma, \perp)$ *has no null planes.*

Proof. Suppose there is a null plane α. Let β be any plane in Σ that intersects α in a line L. Since the space is nonsingular, Theorem 4.3.5 implies that β is a nonsingular plane. But β has a null line L, and so must be isotropic. Taking a point p on L, it follows from Corollary 2.5.4 that in β there is a second null line L' through p (this uses the assumption than the characteristic is not 2).

Now let M be a *non-null* line through p in β, and let N be any (null) line in α distinct from L. By the same reasoning, the plane γ containing M and N is isotropic, and has a second null line N' through p (Figure 4.8).

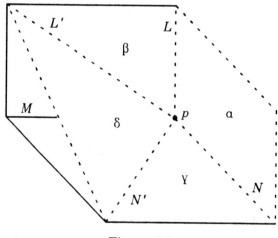

Figure 4.8

The plane δ containing L' and N' then intersects α in a line K which is necessarily null. From the construction it is apparent that $L', N',$ and K are distinct, so in δ there are three null lines through p. But this is only possible if δ is a null plane.

Thus the assumption that there is a null plane implies that there are intersecting null planes (α and δ), which, by Theorem 4.3.5, contradicts the nonsingularity of $(\Sigma, \perp)$.   □

It is noteworthy that this result does not apply in higher dimensions: §5.1 contains an example of a nonsingular metric fourfold that *does* have null planes.

We take up now the question of finding a symmetric matrix over F that characterises the orthogonality relation $\perp$. If o is the point of Σ that serves as the origin when Σ is coordinatised as F^3, then the lines and planes passing through o become the points and lines, respectively, of the projective plane π_F over F (cf. §3.1). In general, a line L will be defined to be *orthogonal to plane* α, $L \perp \alpha$, if L is orthogonal to every line in α. It will be shown now that each line L through o is orthogonal to a unique plane α through o, and that the association of this α with L gives a polarity of π_F. The matrix inducing this polarity (Theorem 3.5.4) is the symmetric matrix we seek.

Theorem 4.4.2. *If L is a non-null line through o, the set $L^\perp$ of all lines through o that are orthogonal to L forms a plane in Σ.*

Proof. That $L^\perp$ forms a plane means that there is a plane α such that $L^\perp$ consists of all and only the lines through o that lie in α, and hence that the points on the lines in $L^\perp$ are precisely all the points in α, so that $L^\perp$ may be identified with α.

First we need two distinct lines M_1, M_2 through o that are orthogonal to L. So, take two distinct planes α_1 and α_2 in Σ that meet along L. Then in each α_i, L is a non-null line, so has a unique altitude M_i through o, with $M_i \neq L$. Thus M_1 and M_2 lie in different planes, and so are distinct as desired.

Now let α be the plane generated by M_1 and M_2. Then by OS4, every line through o in α is in $L^\perp$. Conversely, let $N \in L^\perp$, i.e. N passes through o and $L \perp N$. Then $L \neq N$, since L is non-null.If N did not lie in α, the plane β containing L and N would meet α in a line K distinct from N (Figure 4.9).

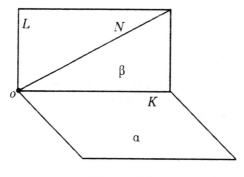

Figure 4.9

But then K is in $L^\perp$, as just noted, and so in β, N and K would be distinct altitudes to L through o, contradicting OS2. Hence N lies in α.

Theorem 4.4.3. *If α is any plane through o, there is exactly one line $\alpha^\perp$ through o that is orthogonal to α.*

Proof. Take first the case that α is a singular plane. Then in α there is a line L through o that is singular in α, so that $L \perp \alpha$. To show that there is no other such line, suppose $M \perp \alpha$, with o on M. Now L is not singular in Σ, so there is some line N not orthogonal to L (clearly N does not lie in α). But $L \perp M$, since $M \perp \alpha$, so M and N are distinct, and generate a plane β meeting α in some line K (Figure 4.10).

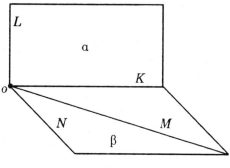

Figure 4.10

Then $L \perp K$, so if $M \neq K$, M and K would generate β, and, by OS4, make $L \perp N$, a contradiction. Therefore $M = K$, so M lies in α, and L and M are intersecting singular lines in α, whence $L = M$ (2.4.2).

Next, take the alternative case that α is not a singular plane. Since there are no null planes (4.4.1), α must be nonsingular, i.e. have no singular lines. Take two distinct non-null lines M, N through o in α (e.g. take a non-null line M and its altitude through o in α). By Theorem 4.4.2, there are planes $M^\perp$ and $N^\perp$ orthogonal to M and N, respectively, and passing through o. Since intersecting planes meet in a line, there is at least one line L through o that lies in both $M^\perp$ and $N^\perp$ (Figure 4.11). Then L is orthogonal to both M and N, so is orthogonal to α by OS4.

Finally, to show that L is unique, suppose that there were a line Q through o with $Q \perp \alpha$ and $Q \neq L$. Then Q and L would generate a plane meeting α in a line R (Figure 4.12). But then if S were any line through o in α, we would have S orthogonal to Q and L, so $S \perp R$. But this makes R a singular line in α, which is impossible, as α is a nonsingular plane.   □

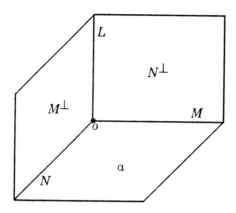

Figure 4.11

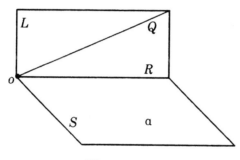

Figure 4.12

Corollary 4.4.4. *Every null line in* $(\Sigma, \perp)$ *lies in a singular plane.*

Proof. Take a null line L through o, and let α be any plane containing L, and $\alpha^{\perp}$ the unique "altitude" to α through o guaranteed by the Theorem. If $L = \alpha^{\perp}$, then L is singular in α as desired. Otherwise, L and $\alpha^{\perp}$ generate a plane β (Figure 4.13). Then β has a null line (L) orthogonal to a line $(\alpha^{\perp})$ that intersects it. This cannot happen in a nonsingular plane, so as β is not a null plane (4.4.1), it must be a singular plane containing L.

Theorem 4.4.5. *If L is any line through o, the set $L^{\perp}$ of lines through o orthogonal to L form a plane.*

Proof. If L is not null, the result is Theorem 4.4.2. If L is null, then by Corollary 4.4.4, there is a plane α containing L as a singular line. Then every line in α through o lies in $L^{\perp}$. Conversely, if M belongs

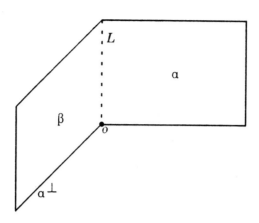

Figure 4.13

to $L^\perp$, M must lie in α, since L is not singular in Σ, and so by Lemma 4.3.3 cannot be orthogonal to any line that does not lie in α.

Theorem 4.4.6.
(1). $(L^\perp)^\perp = L$.
(2). $(\alpha^\perp)^\perp = \alpha$.
(3). *If line L lies in plane α, then plane $L^\perp$ contains line $\alpha^\perp$.*

Proof.
(1). Since $L \perp (L^\perp)$, L must be the unique line $(L^\perp)^\perp$ given by 4.4.3.

(2). If M is the line $\alpha^\perp$, then M is orthogonal to α, so the plane α is included in the plane $M^\perp$. Hence (AS2) $\alpha = M^\perp = (\alpha^\perp)^\perp$.

(3). $\alpha^\perp$ is orthogonal to all lines in α, hence is orthogonal to L, and so lies in $L^\perp$.

Theorem 4.4.7. *The function $f_\perp(L) = L^\perp$ is a polarity of the projective plane π_F.*

Proof. Theorems 4.4.3 and 4.4.5 imply that $f_\perp$ gives a bijective correspondence between lines and planes through o. It follows from 4.4.6(3) that $f_\perp$ is a correlation, i.e. preserves incidence in π_F. For if α is a line in π_F, i.e. an affine plane through o in Σ, then for each affine line L through o in α, the affine plane $f_\perp(L)$ contains $\alpha^\perp$. Thus $f_\perp$ maps the range of projective points L on projective line α to the pencil of projective lines $f_\perp(L)$ passing through the projective point $\alpha^\perp$. In particular, the line-to-point mapping induced in π_F by $f_\perp$ is $f_\perp(\alpha) = \alpha^\perp$. Since then $f_\perp(f_\perp(L)) = L$ (by 4.4.6(1)), $f_\perp$ is involutory.

It remains to prove that $f_\perp$ is a projectivity, and for this it suffices (3.4.1) that the restriction of $f_\perp$ to some projective line β be a projectivity. To show this, choose β to be a *nonsingular* affine plane in Σ (such a plane exists by 4.2.4). Since β is nonsingular, the affine line $f_\perp(\beta) = \beta^\perp$ does not lie in β.

For any line L through o in β, let $g(L)$ be the unique altitude to L through o in β (Figure 4.14).

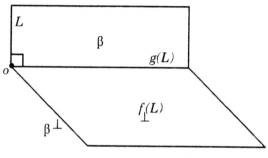

Figure 4.14

Then L is orthogonal to $g(L)$ and to $\beta^\perp$, so $g(L)$ and $\beta^\perp$ both lie in the plane $L^\perp = f_\perp(L)$, and hence generate that plane. In other words, the map $f_\perp$ is the composite $h \circ g$, where $h(K)$ is the plane generated by K and $\beta^\perp$.

Now in π_F, h is the projection of the range of points on β from the point $\beta^\perp$ (Figure 4.15), so if g is a projectivity on β, then $h \circ g = f_\perp$ will be a projectivity.

But g *is* a projectivity! The demonstration of this is given by the Involution Theorem of §3.2. For, if β is realised as the line at infinity of a suitably chosen affine plane γ in Σ, then g is just the involution determined by the orthogonality relation $\perp$ between lines in γ. To spell this out, take γ as any plane of Σ that is parallel to β (think, if you like, of β as the x-y-plane and γ as the plane $z = 1$, as in the discussion of coordinates in §3.1). Now the affine plane γ is embedded in the projective plane π_F of lines and planes through o, by identifying each point on γ with the line joining it to o, and identifying each line L in γ with the unique plane α_L through o that contains L. This plane α_L meets β in a line L' through o parallel to L (Figure 4.16).

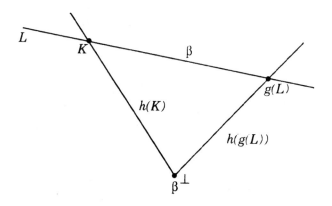

Figure 4.15

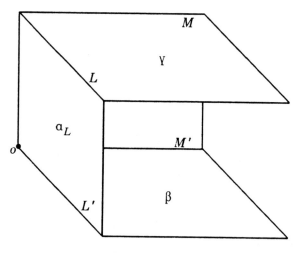

Figure 4.16

By these identifications, every line in π_F, except β, is of the form α_L and corresponds to a line L in γ, with the projective point L' becoming the point at infinity of the affine line L. In other words, the affine plane $\alpha(\pi_F - \beta)$, got by deleting β from π_F, is identified with γ, and inherits the orthogonality relation of the latter, by putting

$$\alpha_L \perp \alpha_M \quad \text{iff} \quad L \perp M.$$

But $L \perp M$ in γ implies $L' \perp M'$ in β (by two applications of OS4), and so $g(L') = M'$. This confirms that g is indeed the projectivity on β induced by the relation $\perp$ in γ (or $\alpha(\pi_F - \beta)$). $\qquad \square$

It follows from the work of §3.5 that there is an invertible symmetric matrix G such that $f_\perp$ is the map f_G induced by G. Now take points a and b (other than o) in Σ, with coordinates (x_1, y_1, z_1) and (x_2, y_2, z_2) respectively. Then $oa \perp ob$ if, and only if, oa lies in the plane $(ob)^\perp = f_G(ob)$. Hence $oa \perp ob$ if, and only if, projective point oa lies on projective line $f_G(ob)$.

Now since ob as a projective point has homogeneous coordinates $[x_2, y_2, z_2]$, $f_G(ob)$ as a projective line has homogeneous coordinates $<l, m, n>$, where

$$(l \ m \ n)^T = Gb^T.$$

Thus projective point $oa = [x_1, y_1, z_1]$ is incident with $f_G(ob)$ if, and only if,

$$x_1 \cdot l + y_1 \cdot m + z_1 \cdot n = 0,$$

i.e. if, and only if,

$$aGb^T = 0.$$

Putting this together gives

$$oa \perp ob \quad \text{iff} \quad aGb^T = 0,$$

and completes the coordinatisation of the nonsingular metric three-fold $(\Sigma, \perp)$.

Self-Polar Triangle

Any polarity in a projective plane has a self-polar triangle that can be used to obtain a coordinatisation of the polarity by a *diagonal* matrix. This was discussed at the end of §3.5, where the existence of the triangle was shown to follow from the necessary existence of a line that is not self-conjugate under the polarity. In the present context, a non-self-conjugate line in π_F under $f_\perp$ is a nonsingular plane through o in $(\Sigma, \perp)$, and the existence of such a plane (Theorem 4.2.4) was used in the proof that $f_\perp$ is a projective correlation.

Let, then, α be a nonsingular plane through o in Σ. A self-polar triangle in π_F arises by taking a non-null line M through o in α, its unique altitude N through o in α, and letting L be the line $\alpha^\perp$ through o orthogonal to α. M and N are distinct, and L does not lie in α, so L and N generate a plane $\beta \neq \alpha$, and L and M generate a third plane γ, with $L \perp \alpha$, $M \perp \gamma$, and $N \perp \beta$.

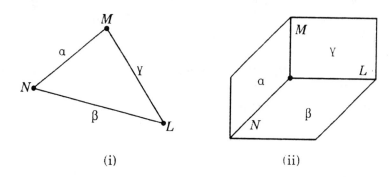

Figure 4.17

In π_F the picture is as in Figure 4.17(i), while the affine perspective is given in Figure 4.17(ii), with L, M, and N appearing as "orthogonal axes". Indeed, by taking L, M, N as the x, y, and z axes, respectively, in coordinatising Σ as F^3, in π_F we get the projective points

$$L = [1, 0, 0], \quad M = [0, 1, 0] \quad N = [0, 0, 1],$$

and the projective lines

$$\alpha = <1, 0, 0>, \quad \beta = <0, 1, 0>, \quad \gamma = <0, 0, 1>,$$

(cf. Figure 3.25). Then, as in the proof of Theorem 3.5.5, this forces the matrix G inducing $f_\perp$ to have the diagonal form

$$\begin{pmatrix} a & 0 & 0 \\ 0 & b & 0 \\ 0 & 0 & c \end{pmatrix}.$$

Suppose now that the coordinatising field F is quadratic, i.e. for each $x \in F$, either x or $-x$ has a square root in F. Then a further simplification of the form of G is possible, showing that there are essentially only two nonsingular geometries over F^3.

Case 1. *All three of the diagonal entries have square roots.*
In this case we recoordinatise Σ, by changing the coordinates (x, y, z) of a point to

$$(x', y', z') = (\sqrt{a} \cdot x, \sqrt{b} \cdot y, \sqrt{c} \cdot z).$$

This transformation is induced by the matrix

$$H = \begin{pmatrix} \sqrt{a} & 0 & 0 \\ 0 & \sqrt{b} & 0 \\ 0 & 0 & \sqrt{c} \end{pmatrix},$$

which is invertible, since G is, and gives a nonsingular linear transformation of F^3.

But if two points have old coordinates (x_i, y_i, z_i), for $i = 1, 2$, then their inner product, as induced by G, is

$$a \cdot x_1 \cdot x_2 + b \cdot y_1 \cdot y_2 + c \cdot z_1 \cdot z_2$$
$$= x_1' \cdot x_2' + y_1' \cdot y_2' + z_1' \cdot z_2',$$

so that in terms of the new coordinates, the inner product is induced by the identity matrix I_3.

Thus in this case, $(\Sigma, \perp)$ can be coordinatised by the Euclidean inner product on F^3.

Case 2. *Two of the diagonal entries have square roots.*
Suppose the two entries in question are a and b. Then since F is quadratic, $-c$ has a square root $\sqrt{-c}$, which we put in place of $\sqrt{c}$ in the recoordinatising matrix H. In terms of this new coordinatisation,

$$a \cdot x_1 \cdot x_2 + b \cdot y_1 \cdot y_2 + c \cdot z_1 \cdot z_2$$
$$= x_1' \cdot x_2' + y_1' \cdot y_2' - z_1' \cdot z_2',$$

giving a characterisation of $\perp$ by the matrix

$$\begin{pmatrix} 1 & 0 & 0 \\ 0 & 1 & 0 \\ 0 & 0 & -1 \end{pmatrix}$$

which induces the Minkowskian inner product on F^3.

Note that under this case it can always be arranged that the last diagonal entry be the one to lack a square root, by suitably interchanging the axes in the coordinatisation of Σ. Thus if the second entry lacked a square root initially, by recoordinatising (x, y, z) as (x, z, y), and interchanging b and c in G, a coordinatisation characterising $\perp$ would result, to which the Case 2 construction applies.

All other cases now reduce to these two. If exactly one diagonal entry has a square root, replace G by $-G$, which defines the same orthogonality relation as G, and so still characterises $\perp$, but has two diagonal entries with square roots (Case 2). Finally, if no entries have square roots, $-G$ falls under Case 1.

The Two Real Cases

The analysis just given shows that over a quadratic coordinatising field (such as **R** or **C**), there are at most two nonsingular metric affine threefolds: those characterised by the Euclidean and the Minkowskian inner products. If $\sqrt{-1}$ exists in F, there is actually only one geometry, for the coordinate change

$$(x, y, z) \mapsto (x, y, \sqrt{-1} \cdot z)$$

transforms the Euclidean inner product into the Minkowskian one.

Over the real-number field **R**, the two cases are easily distinguished in terms of $\perp$, since spacetime has null lines, but Euclidean space does not. Thus by introducing further axioms of order that constrain the coordinatising field to be isomorphic to **R**, we finally arrive at a categorical axiomatisation of the two real cases.

The axioms B1 - B4 of §2.7 for a linear betweenness relation can be retained for affine spaces, with the minor modification that Pasch's Law B4 is relativised to each plane, and becomes the statement

B4'. *If a line L lies in the plane determined by a triangle abc, and passes between a and b but not through c, then L passes between a and c, or between b and c.*

An affine space Σ is *ordered* if there is a ternary relation between points of Σ that satisfies B1 - B3, and B4'. If the continuity axiom B5 of §2.7 is satisfied as well, then Σ is *continuously ordered*. In a continuously ordered Pappian affine space, the coordinatising field satisfies the Dedekind continuity axiom and is isomorphic to **R**.

In view of the discussion of spaces over quadratic fields, we thus have:

Theorem 4.4.8. *A continuously ordered nonsingular metric affine threefold is*

(1) *isomorphic to three-dimensional Minkowskian spacetime if it has at least one null line, and*

(2) *isomorphic to Euclidean three-space otherwise.*

5

Fourfolds

The end goal of this chapter is to obtain categorical axiomatisations of the nonsingular metric affine spaces over $\mathbf{R}^4$. There are three of these: Minkowskian spacetime, Euclidean four-space, and a new geometry whose description we take up immediately.

5.1 Artinian Four-Space
The matrix

$$\begin{pmatrix} 1 & 0 & 0 & 0 \\ 0 & 1 & 0 & 0 \\ 0 & 0 & -1 & 0 \\ 0 & 0 & 0 & -1 \end{pmatrix}$$

induces a nonsingular metric geometry over $\mathbf{R}^4$ which is known as *Artinian four-space*. Its lightcone has equation

$$x^2 + y^2 - z^2 - t^2 = 0,$$

and so intersects the threefold $t = 1$ in the surface S with equation

$$x^2 + y^2 - z^2 = 1.$$

S is a hyperboloid of one sheet (Figure 5.1): in $\mathbf{R}^3$ the equation for S defines the surface generated by rotating the hyperbola $y^2 - z^2 = 1$ from the y-z-plane about the z-axis.

S is actually a *ruled surface*: there are two families of straight lines lying in S. Each family is made up of mutually skew lines, and is known as a *regulus* (Struik [1953], §9.5, Veblen and Young [1910], §103). Through each point on S there passes one line from each regulus, as indicated in Figure 5.1.

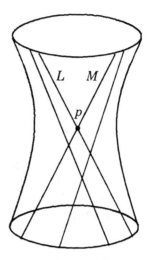

Figure 5.1

Now a point p on S represents a null line through the origin in Artinian four-space. Then the two lines L and M in S through p represent planes in Artinian four-space made up of null lines. Hence the plane in the threefold $t = 1$ that contains the lines L and M represents a Type 2 singular threefold (Figure 4.4) in Artinian four-space, with p representing the singular line through the origin in that threefold. (Note that in Minkowskian spacetime there are no null planes, and hence no Type 2 singular threefolds: cf. Figure 1.17.)

Artinian four-space can be realised as the "orthogonal sum" of two copies of the Lorentz plane (Snapper and Troyer [1971], §36). The notion of orthogonal sum of isotropic planes is in fact crucial to the general classification of metric vector spaces, as the reader may learn from the reference just quoted, and from Artin [1957].

5.2 Affine Fourfolds and Projective Three-Space

In an affine space Σ, a *fourfold* is a subspace of Σ that is generated by a set of five points that do not all lie in any one threefold of Σ. Thus Σ can itself be constrained to be four-dimensional by requiring it to satisfy the axiom:

there exist five points, not all in the same threefold, that generate $\mathcal{P}$.

If Σ is Pappian, and coordinatised as the vector space over some field F, this axiom will entail that the vector space has a four-element basis, and so be isomorphic to the canonical four-dimensional vector space F^4.

Some of the incidence properties of an affine fourfold Σ that we will need to use are the following.

(1). If three non-collinear points lie in a threefold, then every point in the plane they determine (AS2) lies in that threefold.

(2). Any four non-coplanar points lie in exactly one threefold. Hence there is a unique threefold containing (i) a given plane and a point not in that plane, or (ii) a given plane and a line intersecting that plane in a point, or (iii) three concurrent non-coplanar lines, or (iv) two planes that intersect in a line.

(3). If a plane and a threefold intersect, then their intersection is a line.

(4). If two threefolds intersect, then their intersection is a plane.

Some of these properties are a little hard to "see" with our three-dimensional visual imaginations, but with a little practise at proving them one soon gets the "feel" of four-dimensional geometry. Algebraically, some of the properties can be confirmed by using the fact that the general threefold is the solution set of an inhomogeneous linear equation

$$a \cdot x + b \cdot y + c \cdot z + d \cdot t + e = 0$$

with coefficients from F. By the elementary theory of linear equations, the solution set of a pair of such equations, representing two threefolds, is in general two-dimensional, i.e. a plane, confirming (4). Thus a plane can be represented by a pair of linear equations, and so the intersection of a plane and a threefold is the solution set of three equations in four unknowns. Such a solution set is one-dimensional (a line). Similarly, the intersection of two planes is the solution set of four equations in four unknowns. But such a system may have a unique solution! A geometric demonstration of this is given in

Theorem 5.2.1. *If a line L meets a plane α in a single point p, then there is a plane β containing L such that β and α have only the point p in common.*

Proof. Take points q, r in α, and s on L, all distinct from p and each other. Let Γ be the threefold containing p, q, r, s. Then there is a point t not in Γ, as Σ is a fourfold. Let β be the plane containing p, s, and t (Figure 5.2).

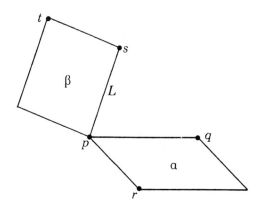

Figure 5.2

Now suppose that β had a point $x \neq p$ in common with α. Then, by hypothesis, x would not be on L, so β would have the non-collinear points p, s, x in common with Γ. By the uniqueness of the plane generated by a non-collinear triple (cf. item (1) above), this would force β to be contained in Γ, contradicting the fact that t is in β but not in Γ.

Projective Spaces

Whereas in F^3 the bundle of lines and planes through the origin forms a two-dimensional projective plane π_F, in F^4 this bundle forms a three-dimensional projective *space*. Abstractly, an incidence structure is defined to be a *projective space* if it satisfies the following axioms.

PS1. *Any two distinct points lie on exactly one line.*

PS2. *If a, b, c, d are four distinct points such that lines ab and cd intersect, then ac and bd also intersect (Figure 5.3).*

PS3. *There exist at least three points on each line.*

Any projective plane satisfies these axioms (as does a structure consisting of a single *projective line*). Notice that it is no longer required that any two lines meet, as in a projective plane. The intention is rather that any two lines *in the same plane* should meet. Hence lines that do not meet are called *skew*, since they can never be parallel in the affine sense of being coplanar but not intersecting.

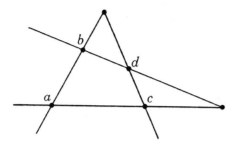

Figure 5.3

The construction of planes, threefolds etc. is simpler in the projective context, because one does not have to take account of parallelism. First, a *subspace* is defined to be any set of points closed under the generation of lines, i.e. if points p and q are in the set, then all points on the line pq are too. Then if S is any set, and p a point not in S, the *subspace generated by p and S* is the set

$$S' = \{q : \text{for some } r \in S, \ q \text{ is on } pr\}$$

of all points that lie on some line through p that intersects S (Figure 5.4).

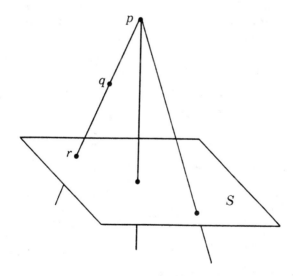

Figure 5.4

Axiom PS2 is needed to prove that S' is indeed a subspace. It is then the smallest subspace containing S and p.

A *plane* in a projective space is defined to be any subspace generated by a line L and a point p not on L (Figure 5.5).

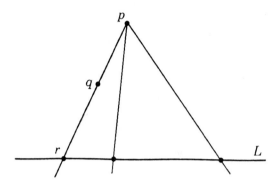

Figure 5.5

Equivalently, a plane is generated by three non-collinear points (any two lines in such a plane must intersect, and so the planes in a projective space are projective planes as defined in §3.1). A *three-space* is a subspace generated by a plane and a point not in that plane, or, equivalently, by four non-coplanar points.

Thus a *projective three-space* can be defined as a projective space which is generated by four non-coplanar points. The set of (affine) lines and planes through the origin in the vector space F^4 is a projective three-space, to be denoted $\pi_3(F)$, whose planes correspond to the *affine threefolds* through the origin in F^4.

Theorem 5.2.2. *In a projective three-space:*
(1). *If line L does not lie in plane π, then L intersects π in a single point.*
(2). *Any two distinct planes intersect in a line.*

Proof. Exercise (using especially PS2). cf. Veblen and Young, §9.

□

Details of the incidence properties of projective spaces may be found in Garner [1981], Chapter 6; Mihalek [1972], Chapter 10; Seidenberg [1962], Chapter 5; and Veblen and Young [1910], Chapter 1.

Projectivities and Collineations
The theory of projectivities in projective three-spaces is broader than the planar case described in §3.2, due to the presence of a

larger variety of "primitive geometric forms" (Veblen and Young, §21). There are three types of *one-dimensional* primitive form:

(1) the range [L] of collinear points that lie on a line L;

(2) the *flat* pencil [p] of concurrent lines that pass through a point p in a given plane; and

(3) the pencil {L} of *coaxial planes* that contain a line L.

Associated with these forms are three types of projection, and three types of section.

(1). The projection [L] ▷ {M} of a range [L] of points from a line M that does not meet L: this assigns to each point p on L the unique plane containing p and M (Figure 5.6).

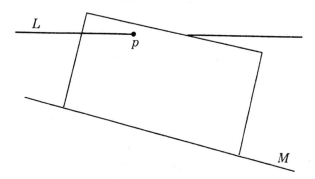

Figure 5.6

Inversely, the section {M} ◁ [L] of a coaxial pencil {M} of planes by a line L skew to M assigns to each plane containing M its point of intersection with L (which always exists, by Theorem 5.2.2(1), if L and M do not meet).

(2). The projection [L] ▷ [p] of range [L] from point p not on L assigns to each point x on L the line joining x to p. The image of this projection is the flat pencil of all lines through p in the unique plane containing L and p. The inverse of the projection is the section [p] ◁ [L] (recall Figures 3.4 and 3.5). Such a section of a flat pencil [p] by a line L exists if, and only if, the line L lies in the plane of the pencil and does not pass through its centre p.

(3). The projection [p] ▷ {M} of a flat pencil [p] from a line M that passes through its centre p: this assigns to each line L in the pencil the unique plane containing L and M (Figure 5.7).

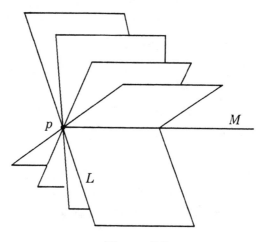

Figure 5.7

Inversely, there is the section $\{M\} \lhd [p]$ of a coaxial pencil $\{M\}$ of planes by a plane that is not in the pencil (i.e. does not contain M), and hence meets M in a point p.

A *projectivity* in a projective three-space is a bijection between two one-dimensional primitive forms that is the composition of a finite number of projections and sections. If the projective space is Pappian (i.e. each plane of the space is Pappian), then the uniqueness of planar projectivities embodied in the Fundamental Theorem (§3.2) extends to the following results.

(1). A projectivity on a range $[L]$ is the identity if it leaves three distinct points of L fixed.

(2). A projectivity between any two one-dimensional forms is uniquely determined by any three elements of its domain and their images.

(cf. Veblen and Young [1910], §§35, 36, 50).

A *collineation* of a projective space is, as for any incidence structure, an isomorphism f from the space to itself. Thus f is a bijection from $\mathcal{P}$ to $\mathcal{P}$ that maps a collinear range $[L]$ onto another collinear range, whose axis is denoted $f(L)$. This extends f to a bijection between lines, with p on L if, and only if, $f(p)$ is on $f(L)$. In a projective space of more than two dimensions, f also lifts to a bijection between planes: if L and M are two lines in a plane π, then the image $f(\pi)$ of π is a plane containing $f(L)$ and $f(M)$. Moreover, f induces an isomorphism between π and $f(\pi)$.

A collineation is *projective* if it induces a projectivity on each line of the space. As in the planar case, we have:

Theorem 5.2.3. *A collineation of a projective three-space is projective if it induces a projectivity on at least one line L.*

Proof. Let f induce a projectivity on L. Take a plane π containing L. Then f induces a projectivity on any line M in π, by essentially the same construction as Theorem 3.4.1, using a perspectivity between L and M in π, and the fact that f induces an isomorphism between π and its image plane $f(\pi)$.

Now let K be any line that does not lie in π. If π' is any plane containing K, then π' meets π in a line M (Theorem 5.2.2(2)). By the previous paragraph, f induces a projectivity on M. Hence as M lies in π', f induces a projectivity on all lines in π', by the same argument as for π. Thus f induces a projectivity on K. $\qquad\square$

Extending the arguments of Theorem 3.4.2 shows that in a Pappian projective three-space:

(1) a projective collineation is the identity function if it fixes a five-point (i.e. five points, no four of which are coplanar); and

(2) a projective collineation is uniquely determined by its action on any five-point (i.e. two such mappings that agree on some five-point must be identical).

(Veblen and Young, §35).

Correlations and Polarities

In a projective plane, the principle of duality allows points and lines to be interchanged, and is embodied in the construction of the dual plane, whose points (respectively, lines) are the lines (respectively, points) of the original plane. A similar duality principle holds in three projective dimensions, by interchanging points and *planes*, and leaving lines as they are. The dual space of a given projective three-space has the planes of the given space as its points, and the lines of the given space as its lines. This dual structure is a projective three-space itself: for instance, the fact that two planes always intersect in a line (5.2.2) shows that the dual satisfies PS1. The planes of the dual space correspond to the points of the original space (see two paragraphs below for an indication of how, or Veblen and Young §11).

A *correlation* of a projective three-space is an isomorphism f from that space to its dual space. Thus f may be described as a bijection from the set of points of the space onto its set of planes, mapping collinear points to coaxial planes. The image of the range $[L]$ of all points on line L will be a coaxial pencil $\{M\}$ of planes, whose axis (common line) M will be denoted $f(L)$. This extends f to a bijection $L \mapsto f(L)$ between the lines of the space, with point p on line L if, and only if, plane $f(p)$ contains line $f(L)$.

A correlation f maps three non-collinear points p, q, r in a plane π to three non-coaxial planes $f(p), f(q), f(r)$, meeting pairwise in three non-coplanar lines, which in turn concur at some point s. Then each point in π is mapped by f to a plane through s. In other words, the image of a plane π under f is the bundle of all planes through some point s. Writing $f(\pi)$ for this point s extends f to a bijection from planes to points. Hence it makes sense to talk about inverses and compositions of correlations, as in the planar case.

A correlation is *projective* if it induces on any line L a projectivity between the range of points on L and the pencil of planes containing the line $f(L)$. Dualising the argument for collineations (Theorem 5.2.3) shows:

Theorem 5.2.4. *A correlation of a projective three-space is projective if it induces a projectivity on at least one line.* $\square$

The inverse of any projective correlation is also a projective correlation, while the composition of two projective correlations is a projective *collineation*. By the uniqueness properties of the latter, it follows that:

Theorem 5.2.5. *In a Pappian projective three-space, a projective correlation is uniquely determined by its action on any five-point.*

$\square$

A *polarity* is a projective correlation f that is equal to its own inverse. This means that the inverse of the plane-to-point mapping induced by f is just f itself, so that if $f(p) = \pi$, then $f(\pi) = p$. The effect of a polarity is thus to interchange points and planes. $f(p)$ is the *polar plane* of point p, while the point $f(\pi)$ is the *pole* of plane π.

Matrix-Induced Polarities

In the projective three-space $\pi_3(F)$ of all (affine) lines through the origin in the vector space F^4, projective points and planes can be given homogeneous coordinates, as follows.

(1). A point p is a proportionality class $[x_1, x_2, x_3, x_4]$ of non-zero quadruples of elements of F (each quadruple being a point (other than **o**) on p, when the latter is viewed as a line through **o** in F^4).

(2). A plane π is a proportionality class $< a_1, a_2, a_3, a_4 >$ of non-zero quadruples of elements of F, with the point p of (1) lying in π iff

$$a_1 \cdot x_1 + a_2 \cdot x_2 + a_3 \cdot x_3 + a_4 \cdot x_4 = 0$$

(this being the linear equation whose solution set is π when the latter is viewed as an affine three-fold passing through **o** in F^4).

A 4×4 invertible matrix A over F induces a bijection f_A from points to planes of $\pi_3(F)$, by putting

$$f_A([x_1, \ldots, x_4]) = < a_1, \ldots, a_4 >,$$

where

$$(a_1 \ldots a_4)^T = A(x_1 \ldots x_4)^T.$$

Using the five-point

$$e_1 = [1, 0, 0, 0], \quad e_2 = [0, 1, 0, 0], \quad e_3 = [0, 0, 1, 0],$$
$$e_4 = [0, 0, 0, 1], \quad u = [1, 1, 1, 1],$$

and arguments similar to those developed in Chapter 3, the following are shown.

(1). f_A is a projective correlation.

(2). There is a unique matrix induced transformation mapping any given five-point onto any given five-plane (a five-plane consists of five planes, no four of which are on a common point).

Hence, by the uniqueness of projective correlations (5.2.5), we get

Theorem 5.2.6. *Any projective correlation of $\pi_3(F)$ is the correlation f_A induced by some invertible 4×4 matrix A.*   □

If correlation f_A is a polarity, then $A^T = \pm A$ (cf. the proof of Theorem 3.5.4). But if $A^T = -A$, i.e. A is skew-symmetric, then $vAv^T = 0$ for any vector v in F^4, so that each point of $\pi_3(F)$ lies on its polar plane. In that case f_A is called a *null polarity* (Coxeter [1942], p.92, Struik [1953], p.250), something that cannot occur in a projective plane, but is possible in three-space.

The type of orthogonality relation on F^4 being considered in this chapter does not generate a null polarity on $\pi_3(F)$ (and of course a skew-symmetric matrix would induce a bilinear form on F^4 satisfying $a \bullet b = -b \bullet a$, instead of $a \bullet b = b \bullet a$). To obtain a symmetric matrix inducing a correlation it is necessary to consider the notion of a *self-polar tetrahedron*.

A tetrahedron is the figure formed by four non-coplanar points p, q, r, s in $\pi_3(F)$ (Figure 5.8).

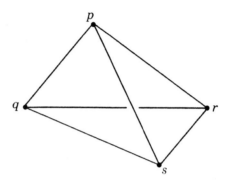

Figure 5.8

Any three of these points are non-collinear, and lie in a plane which is a *face* of the tetrahedron. The fourth point is not in this plane, and is the *opposite vertex* of the face. A tetrahedron is *self-polar* with respect to a correlation f if each face is the polar plane under f of its opposite vertex.

Theorem 5.2.7. *If a projective correlation f of $\pi_3(F)$ has a self-polar tetrahedron, then there is a coordinatisation of $\pi_3(F)$ relative to which f is induced by some diagonal matrix.*

Proof. A tetrahedron in $\pi_3(F)$ consists of four lines through **o** in F^4 that do not all lie in the same three-dimensional subspace of F^4. Hence (by recoordinatising) we can assume that these lines are the coordinate axes of F^4, so that in $\pi_3(F)$ we get

$$p = [1,0,0,0], \quad q = [0,1,0,0], \quad r = [0,0,1,0], \text{ and } s = [0,0,0,1].$$

Then the face opposite p has homogeneous plane-coordinate

$$<1,0,0,0>,$$

and similarly for the other vertices (cf. Figure 3.25 in the proof of Theorem 3.5.5). By the previous result, f is f_A for some invertible matrix A. But as

$$f_A([1, 0, 0, 0]) = \; <1, 0, 0, 0>,$$

and similarly for the other vertices, A must be diagonal.   □

To round out this discussion of projective three-spaces, it may be observed that the description of coordinates for the space $\pi_3(F)$ of lines through the origin in F^4 suggests a method for coordinatising an abstract Pappian projective three-space. Pick a plane π in such a space as the "plane at infinity", and delete all points on π. Since any two planes in the space meet in a line, deletion of π removes one line from each of the other planes, and turns them into affine planes. Overall what remains is a Pappian affine threefold, which can be coordinatised as F^3 for some field F. Then F^3 can be identified with $F^3 \times \{1\}$, i.e. with the threefold in F^4 defined by the equation $x_4 = 1$. Equivalently, the points of F^3 can be identified with the lines through o in F^4 that meet the threefold $x_4 = 1$, i.e. with those lines that are not in the threefold $x_4 = 0$. Hence points not in π in the original projective space get homogeneous coordinates $[x_1, x_2, x_3, x_4]$ with $x_4 \neq 0$. Points in π correspond to lines through o in F^4 that do not meet the threefold $x_4 = 1$. Such points get coordinates of the form $[x_1, x_2, x_3, 0]$.

Since this general construction will not be needed, its finer points are left to the reader.

5.3 Nonsingular Fourfolds

Let $(\Sigma, \perp)$ be a nonsingular metric affine fourfold. Σ is an affine fourfold (§5.2), coordinatised as the vector space F^4 over some field F, which we continue to assume is not of characteristic 2. $\perp$ is a relation between lines of Σ satisfying axioms OS1 - OS5 of §4.2, and having no lines that are singular in Σ.

The main conclusion of this section will be that there is a 4×4 diagonal matrix over F characterising $\perp$.

Lemma 5.3.1. If Γ is an affine threefold in Σ, and line L is singular in Γ, then L is not orthogonal to any line that does not lie in Γ.

Proof. Let p be any point on L. Suppose p is orthogonal to some line not in Γ. Then there is a line M through p with $L \perp M$ and M not lying in Γ (by OS5 etc.). Let N be any line through p in Σ. If N is L, M, or any line in Γ, then $L \perp N$. Otherwise, M and N generate a plane α that intersects Γ in a line $K \neq M$ (Figure 5.9).

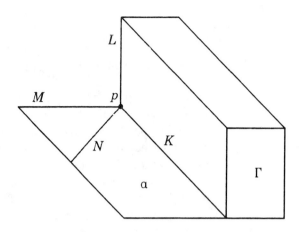

Figure 5.9

Then $L \perp M$ and $L \perp K$, so by OS4, $L \perp N$.

But this implies that L is singular in Σ, a contradiction.

Theorem 5.3.2.

(1). Σ *has no null threefolds.*

(2). Σ *has no singular threefolds of Type 1.*

Proof.

(1). Suppose that Γ is a null threefold in Σ. Since Σ has nonsingular planes through each point (Theorem 4.2.4), and all planes of Γ are null, there must be a nonsingular plane β in Σ that intersects Γ in a (null) line L. Pick a point p on L.

Now perform a construction similar to that of Figure 4.8 in the proof of Theorem 4.4.1 (Figure 5.10). First, β is a nonsingular plane with null line L, so is isotropic and has another null line L' through p in β. Also β has a line M through p that is not null, and hence does not lie in Γ. Next, take a line N through p in Γ with $N \neq L$. Let γ be the plane generated by M and N. Now N is singular in Γ (as Γ is a null space), so by Lemma 5.3.1 it cannot be that $N \perp M$. Thus plane γ has a nonsingular null line N, and hence has a second null line N' through p. The plane δ generated by L' and N' then meets Γ in a null line K distinct from L' and N'. Hence there are three null lines through p in δ, forcing δ to be a null plane. But then $K \perp L'$ in δ, which is a contradiction, by Lemma 5.3.1, since K is singular in Γ and L' does not lie in Γ.

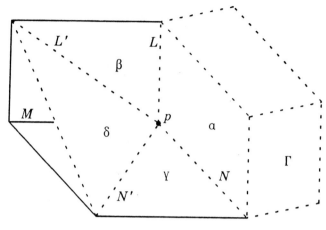

Figure 5.10

(2). Suppose that Γ is a singular threefold in Σ that is of Type 1, i.e. has a pair of intersecting singular lines. Then, as described in §4.3 (Theorem 4.3.1), Γ has a null plane α whose lines are all singular in Γ.

Let Δ be a threefold that intersects Γ in the plane α (i.e. take a point not in Γ and let Δ be the threefold generated by this point and α). Then Δ contains the null plane α, and so is singular (Theorem 4.4.1). Take a point p in α, and let L be a line that is singular in Δ and passes through p (Figure 5.11).

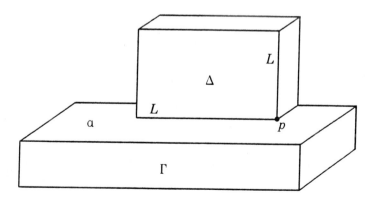

Figure 5.11

Now if L lies in α, then L is singular in Γ and orthogonal as well to all lines in Δ, which contradicts Lemma 5.3.1. On the other hand, if L does not lie in α, then any line that does lie in α is singular in Γ and orthogonal to the line L which does not lie in Γ. Again this contradicts Lemma 5.3.1.

Hence such a threefold Γ cannot exist.

Theorem 5.3.3. *Let $M, N,$ and K be three non-coplanar lines in Σ concurring at a point p. If a line L through p is orthogonal to each of $M, N,$ and K, then L is orthogonal to every line through p that lies in the threefold generated by $M, N,$ and K.*

Proof. Let Γ be the threefold generated by $M, N,$ and K. If Q is any line through p in Γ with $Q \neq K$, then Q and K generate a plane α in Γ. This plane intersects the plane β generated by M and N in a line R (Figure 5.12).

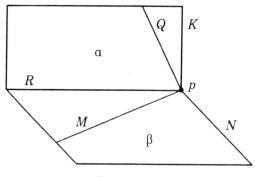

Figure 5.12

Since L is orthogonal to M and N, axiom OS4 implies that L is orthogonal to all lines in β, hence in particular $L \perp R$. But then $L \perp K$ and $L \perp R$, so again by OS4, L is orthogonal to all lines in α, and hence $L \perp Q$.   □

We now proceed to construct the polarity of $\pi_3(F)$ induced by the orthogonality relation $\perp$ in Σ. Let o be the point of Σ coordinatised as the origin in F^4.

Theorem 5.3.4. *If L is a non-null line through o, the set $L^\perp$ of all lines through o that are orthogonal to L forms a threefold in Σ.*

Proof. As in the proof of Theorem 4.4.2, by taking two distinct planes containing L, the altitudes to L through o in each plane give us two distinct lines M and N through o such that L does not lie in the plane α generated by M and N (Figure 5.13).

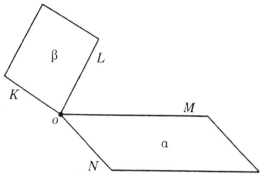

Figure 5.13

By Theorem 5.2.1, there is a plane β containing L and meeting α only at o. Let K be the altitude to L through o in β.

Now since M, N, and K are not coplanar, they generate a threefold Γ. By the previous Theorem, L is orthogonal to every line through o in Γ, i.e. every such line is in $L^\perp$. But if Q is any line in $L^\perp$, then Q must lie in Γ. For otherwise, L and Q would generate a plane meeting Γ in a line $R \neq Q$ (Figure 5.14), and so in this plane L would have distinct altitudes Q and R through p, contrary to OS2.

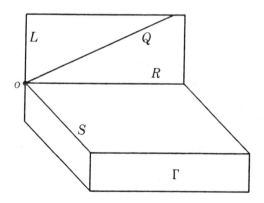

Figure 5.14

Theorem 5.3.5. *If Γ is any threefold through o in Σ, then there is exactly one line through o that is orthogonal to Γ (i.e. orthogonal to all lines through o in Γ).*

Proof. Take first the case that Γ is a singular threefold. Then there is a line L through o in Γ that is singular in Γ, and hence orthogonal to Γ. To show that L is the only such line, suppose that a line M through o is orthogonal to Γ. Then in particular $M \perp L$, and so by Lemma 5.3.1 M must lie in Γ. Hence M is also a singular line through o in Γ. But Γ is not a Type 1 threefold (Theorem 5.3.2), and so has no intersecting singular lines (§4.3). Therefore $L = M$.

Next, as Γ has no null threefolds (5.3.2), the only other case is that Γ is nonsingular. Let M, N, and K be three non-null lines through o in Γ that are non-coplanar, and hence generate Γ (such lines do exist: by the self-polar triangle construction of §4.4 they can even be chosen mutually orthogonal if desired). By Theorem 5.3.4 there are threefolds $M^\perp, N^\perp, K^\perp$ through o that are orthogonal to M, N, and K respectively. Since two intersecting threefolds meet in a plane in Σ, and a plane intersects a threefold in a line, $M^\perp, N^\perp$, and $K^\perp$ have at least one line L in common through o. Then L is orthogonal to each of M, N, and K, so is orthogonal to Γ (Theorem 5.3.3).

Finally, if Q is any line through o orthogonal to Γ, it must be that $Q = L$. For otherwise, L and Q generate a plane having a line R in common with Γ. Then if S is any line in Γ (Figure 5.14), S is orthogonal to L and to Q, hence orthogonal to R. But this makes R a singular line in Γ, contradicting the nonsingularity of the latter.

Corollary 5.3.6. *There exist nonsingular threefolds through o in Σ.*

Proof. Σ is not a null space, and so has a non-null line L through o. Let Γ be the threefold $L^\perp$ given by Theorem 5.3.4. Then Γ is nonsingular, for otherwise there would be a singular line M through o in Γ. Singular lines being null, L and M would then be distinct lines through o orthogonal to Γ, contradicting Theorem 5.3.5. $\square$

Our next task is to extend Theorem 5.3.4 to show that a null line is also orthogonal to some threefold (in which it will be singular). The situation is more involved than the three-dimensional analogue (4.4.4: every null line lies in a singular plane), because of the possible presence of Type 2 singular threefolds containing nonsingular null lines. Dealing with this case involves, in effect, demonstrating that if Σ contains a Type 2 threefold, then it has the property that is characteristic of Artinian four-space, namely that of being the "orthogonal sum" of two Artinian (isotropic) planes.

Theorem 5.3.7. *If L is any null line through o in Σ, there is a threefold in which L is singular.*

Proof. Let Γ be any threefold containing L.

Case 1: Γ is nonsingular. Then there is a singular plane α in Γ containing L (Corollary 4.4.4), with L of course singular in α. Let M be any line through o in α with $M \neq L$, and let N be the unique line through o in Σ that is orthogonal to Γ (Theorem 5.3.5). Then N does not lie in Γ (Figure 5.15), as Γ is nonsingular.

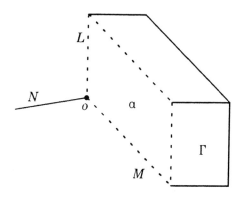

Figure 5.15

Hence N is not coplanar with L and M, so these three lines generate a threefold Δ. But L is orthogonal to L, M, and N, so is orthogonal to all lines through o in Δ, by Theorem 5.3.3. Thus L is singular in Δ.

Case 2: Γ is a Type 3 singular threefold. Then there is a singular line through o in Γ that is intersected only by non-null lines. Since L is null, it must be this singular line.

Case 3: Γ is a Type 2 singular threefold. If L is the singular line in Γ through o, the proof is finished. Otherwise, L is nonsingular in Γ, and so must be nonsingular in some plane α in Γ. Let M and N be distinct non-null lines through o in α (e.g. any non-null line and its altitude). Then the threefolds $M^\perp$ and $N^\perp$ of Theorem 5.3.4 are distinct, or else M and N would be distinct lines through o orthogonal to $M^\perp$, contradicting Theorem 5.3.5. Hence $M^\perp$ and $N^\perp$ intersect in a plane β. Every line through o in β is orthogonal to M and N, hence is orthogonal to α, and in particular to L (Figure 5.16). It follows that L does not lie in β, or else L would be orthogonal to α, contradicting the nonsingularity of L in α. Thus L and β generate a threefold Δ. In Δ, L is orthogonal to L (being null), and to all lines through o in β, and hence L is singular in Δ (5.3.3). $\square$

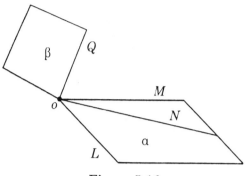

Figure 5.16

Notice that in Case 3 of the proof of this Theorem, the plane α is isotropic, since it has a nonsingular null line L. The reader may enjoy the exercise of proving that β is also isotropic, by showing that β intersects Γ in the unique singular line Q through o in Γ (hence β has only the point o in common with α), and that in β, Q is a nonsingular null line.

Theorem 5.3.8. *If L is any line through o in Σ, the set $L^\perp$ of all lines through o that are orthogonal to L forms a threefold in Σ.*

Proof. If L is non-null, the result is Theorem 5.3.4. If L is null, then by the previous result there is a threefold Γ in which L is singular. Hence all lines through o in Γ are in $L^\perp$. But if M belongs to $L^\perp$, then since $L \perp M$ and L is singular in Γ, M must lie in Γ by Lemma 5.3.1.

Theorem 5.3.9. *If α is any plane through o, the set $\alpha^\perp$ of all lines through o that are orthogonal to α forms a plane. If α is a null plane, then it is equal to $\alpha^\perp$. If α is singular, then it meets $\alpha^\perp$ in a line. If α is nonsingular, then it has only the point o in common with $\alpha^\perp$.*

Proof. The construction has already been used in the proof of Case 3 of Theorem 5.3.7. This time, take any two distinct lines M, N through o in α. The threefolds $M^\perp$ and $N^\perp$ are then distinct, or else M and N would be distinct lines orthogonal to $M^\perp$ (5.3.8, 5.3.5). Hence these threefolds intersect in a plane β. Every line through o in β is orthogonal to M and to N, hence is orthogonal to α, and so lies in $\alpha^\perp$. Conversely, any line in $\alpha^\perp$ lies in both $M^\perp$ and $N^\perp$, hence in β.

Now if α is null, then every line through o in α lies in the plane $\alpha^\perp$. Since a plane is determined by three non-collinear points, this forces $\alpha = \alpha^\perp$.

If α is singular, then the singular line L through o in α lies in $\alpha^\perp$. But then the two planes cannot have any point not on L in common, since this would make $\alpha = \alpha^\perp$ and α a null plane. Thus α and $\alpha^\perp$ meet in a line (like the orthogonal planes $x = 0$ and $y = 0$ in the Robb threefold).

Finally, if α is nonsingular, it cannot have a point other than o in common with $\alpha^\perp$, or else the line joining this point to o would lie in both planes, and hence be singular in α.

Theorem 5.3.10. *The function $f_\perp(L) = L^\perp$ is a projective correlation of the projective three-space $\pi_3(F)$, and has a self-polar tetrahedron.*

Proof. $f_\perp$ is a bijection between affine lines and affine threefolds through o (5.3.5, 5.3.8), i.e. between projective points and projective planes in $\pi_3(F)$. If line L lies in an affine plane α through o, then every line in the plane $\alpha^\perp$ is orthogonal to L, and so lies in $L^\perp$. Hence in $\pi_3(F)$, $f_\perp$ maps the range of points L on the projective line α onto the coaxial pencil of projective planes $L^\perp$ containing the projective line $\alpha^\perp$. Thus $f_\perp$ is a correlation.

To show that $f_\perp$ is projective, it suffices to show that it induces a projectivity on at least one line. For this, let β be a nonsingular plane through o. Then for each line L through o in β, let $g(L)$ be the altitude to L through o in β. Since Σ contains nonsingular threefolds through o (Corollary 5.3.6), we can choose β to be a plane in such a threefold Γ, and show, as in Theorem 4.4.7, that in $\pi_3(F)$, g is a projectivity on the projective line β in the projective plane Γ. Now the affine plane $\beta^\perp$ of Theorem 5.3.9 meets β only at o, since β is nonsingular, and so in $\pi_3(F)$, β and $\beta^\perp$ become skew lines (Figure 5.17). This means that we can form the projection h of the range of points on β from the line $\beta^\perp$. h assigns to each point M on β the projective plane containing M and $\beta^\perp$. From the point of view of Σ, h assigns to the affine line M the unique affine threefold containing M and the affine plane $\beta^\perp$.

But if L lies in β, then L is orthogonal to $g(L)$ and to $\beta^\perp$, so the line $g(L)$ and the plane $\beta^\perp$ both lie in the threefold $L^\perp$, i.e. $L^\perp = h(g(L))$. This shows that $f_\perp$ is the composition of the projectivity g with the projection h, and so is itself a projectivity.

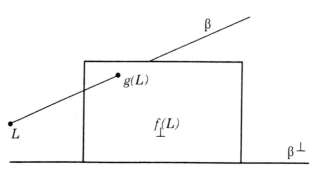

Figure 5.17

Finally, to obtain a self-polar tetrahedron, use the fact that the nonsingular threefold Γ has a self-polar triangle (as shown in §4.4), made up of three mutually orthogonal and non-coplanar affine lines L, M, N through o. Let K be the unique line through o that is orthogonal to Γ (5.3.5), and hence orthogonal to each of L, M, and N. Then K does not lie in Γ, since the latter is nonsingular. Each of the four lines L, M, N, K is orthogonal to the other three, and hence is orthogonal to the threefold generated by the other three. In other words, in $\pi_3(F)$, $f_\perp$ maps each vertex of the tetrahedron $LMNK$ to its opposite face (Figure 5.18). $\square$

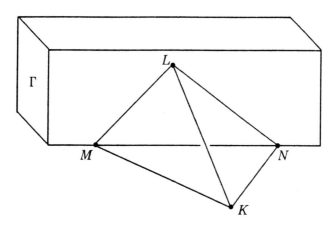

Figure 5.18

The coordinatisation of the nonsingular metric affine fourfold $(\Sigma, \perp)$ is now in place. Theorem 5.2.7 implies that there is a coordinatisation of Σ as F^4 relative to which $f_\perp$ is the transformation induced by some 4×4 invertible diagonal matrix G. Multiplication by G transforms a homogeneous line-coordinate for L to a homogeneous plane-coordinate for $L^\perp$. Then reasoning just as in the three-dimensional case of §4.4, it follows that the inner product on F^4 defined by G characterises $\perp$.

5.4 The Three Real Fourfolds

Let $(\Sigma, \perp)$ be a nonsingular metric affine fourfold, coordinatised by a diagonal matrix

$$\begin{pmatrix} a & 0 & 0 & 0 \\ 0 & b & 0 & 0 \\ 0 & 0 & c & 0 \\ 0 & 0 & 0 & d \end{pmatrix}$$

over a *quadratic* field F. Then, as for the three-dimensional case discussed in §4.4, appropriate recoordinatisations will reduce the diagonal entries to ± 1.

Case 1. a, b, c, d all have square roots. Then the coordinate change

$$(x, y, z, t) \mapsto (\sqrt{a} \cdot x, \sqrt{b} \cdot y, \sqrt{c} \cdot z, \sqrt{d} \cdot t) \tag{†}$$

produces a coordinatisation of Σ in which $\perp$ is characterised by the identity matrix I_4.

Case 2. Three diagonal entries of G have square roots. By first interchanging coordinate axes if necessary, these three can be taken to be a, b, and c. Then replacing d by $-d$ in (†) coordinatises $\perp$ by the Minkowskian inner product induced by matrix

$$\begin{pmatrix} 1 & 0 & 0 & 0 \\ 0 & 1 & 0 & 0 \\ 0 & 0 & 1 & 0 \\ 0 & 0 & 0 & -1 \end{pmatrix}.$$

Case 3. Two diagonal entries have square roots. Taking these as a and b, and replacing c and d by $-c$ and $-d$ in (†) then gives a coordinatisation of $\perp$ by the inner product induced by

$$\begin{pmatrix} 1 & 0 & 0 & 0 \\ 0 & 1 & 0 & 0 \\ 0 & 0 & -1 & 0 \\ 0 & 0 & 0 & -1 \end{pmatrix}.$$

Finally, observe that replacing G by $-G$ reduces the remaining cases (1 or 0 diagonal entries with square roots) to one of the above.

Thus there are at most three nonsingular metric affine fourfolds over a quadratic field F. In the case $F = \mathbf{R}$, the three are distinguishable by simple orthogonality properties of lines: Euclidean four-space has no null lines; Minkowskian spacetime has no null planes, and hence no intersecting orthogonal null lines; while Artinian four-space does have the latter.

Classification Theorem.

A continuously ordered nonsingular metric affine fourfold is

(1) *isomorphic to Euclidean four-space if it has no null lines;*

(2) *isomorphic to Artinian four-space if it has a pair of intersecting null lines that are orthogonal; and*

(3) *isomorphic to Minkowskian spacetime otherwise, i.e. if it has null lines, but no intersecting orthogonal null lines.*

Appendix A

Metageometry

The axiomatic method that is the hallmark of contemporary mathematics stems from Euclid's immortal *Elements*, which stood for two millenia as the paradigm of systematic presentation of deductive reasoning. With the hindsight provided by progress in mathematical logic and foundational studies, Euclid's work is now regarded as having a number of deficiencies. Apart from the use of hidden assumptions, and attempts to define what should be taken as primitive notions, the major shortcoming is the absence of a theory of relative position (order) of points. This lacuna has allowed the production from time to time of such fallacies as Euclidean "proofs" that all triangles are isosceles, and that there exist triangles with two right angles etc.

As mentioned in §2.7, an explicit treatment of order was developed in the latter part of the last century, and the work of a number of mathematicians (Pasch, Peano, Pieri, Hilbert, Veblen et. alia) resulted in the rigorous foundation that we know today. The most well-known formulation is Hilbert's system, dating from about 1899, and based on the undefined notions of *point, line, plane, incidence, betweenness,* and *congruence* of line segments.

Since Hilbert's work there have been few departures in axiomatic geometry of a foundational nature. One that does have considerable conceptual significance is due to Alfred Tarski, and was carried out in 1930 (although not published until much later, cf. Tarski [1951], [1959], [1967]).

Now the Dedekind continuity axiom (§2.7) involves quantifiers ranging over sets of points in an ordered field or on a line (for all non-empty *subsets* C and D ... etc.). This is *second-order* quantification. All other quantifiers in Hilbert's axioms are *first-order*, ranging only over the points (for all points p ..., there exists a point q ... etc.). Tarski's idea was to restrict the Dedekind continuity axiom to apply only to sets C, D that are defined by properties expressible in the *first-order* language of Hilbert's primitives. The resulting system is no longer categorical: it has models other than Euclidean space over **R**. Indeed it has models of every infinite cardinality, including countable models, such as the one based on the field of real algebraic numbers. But in place of categoricity one gets an axiom system that is *deductively complete*, in the sense that each first-order statement is either provable , or refutable (i.e. its negation is provable). The provable statements are precisely those that are true in the geometry over **R**. Moreover there is an algorithm for deciding whether a given statement is provable or not. This is often express by saying that the set of first-order statements is *decidable*.

In this Appendix it will be shown how to adapt the Tarskian approach to give a complete and decidable axiomatisation for Minkowskian spacetime. Some familiarity will need to be assumed with the elements of first-order logic.

Real-Closed Ordered Fields
An ordered field F is *real-closed* if it satisfies

(1) every positive element has a square root, i.e.

$$\forall x(0 < x \rightarrow \exists y(y^2 = x)),   \tag{a}$$

and

(2) every polynomial of odd degree has a zero.

The assertion that all polynomials of degree n have a zero can be expressed by the sentence

$$\forall x_0 \ldots \forall x_n \exists y(x_0 + x_1 \cdot y + \cdots + x_n \cdot y^n = 0)   \tag{b_n}$$

in the first-order language L_{OF} of ordered fields, the latter having the symbols

$$+, \ -, \ \cdot, \ 0, \ 1, \ <$$

(the term y^m is defined inductively as $y \cdot y^{m-1}$, with y^0 being 1).

The *theory* of real-closed ordered fields is the set T_{ROF} of first-order L_{OF}- sentences comprising the ordered-field axioms, the sentence (a), and the sentences (b_n) for *odd* natural numbers n. Instead of the b_n's one could take all universal closures of formulae of the form

$$[(\exists x\varphi) \wedge \exists y \forall x(\varphi \to y < x)] \to \exists y \forall z[y < z \leftrightarrow \exists x(\varphi \wedge (x < z))],$$

where φ is an L_{OF}-formula in which x is free (unquantified), but y and z are not (Tarski [1967], p.15: the universal closure of a formula results by prefixing it with universal quantifiers for all its free variables). Taken together, these sentences amount to a restriction of the assertion "every non-empty set that is bounded below has a greatest lower bound" to sets definable by L_{OF}-formulae.

Note that, since positive elements all have square roots, a real-closed ordered field is *quadratic*.

The models of T_{ROF} are precisely the real-closed ordered fields. Tarski developed an algorithm for showing that T_{ROF} admits *elimination of quantifiers*. To each L_{OF}-sentence σ, the algorithm associates a sentence σ^+ that has no variables, is provable from the axioms T_{ROF} to be equivalent to σ, and hence is true of any real-closed ordered field F if, and only if, σ is true of F. But any sentence without variables is either provable or refutable from T_{ROF} (Tarski [1967], Lemma 2.9), implying the deductive completeness of T_{ROF}. Indeed there is an algorithm for showing that every sentence without variables is equivalent to one of the sentences $(0 = 0)$ and $(0 = 1)$ (Tarski [1951], Theorem 36).

Since T_{ROF} is deductively complete, any L_{OF}-sentence is either true of all T_{ROF}-models (real-closed ordered fields), or false in all such models. The following facts obtain.

- An L_{OF}-sentence is true of the real-number field **R** if, and only if, it is true of all real-closed ordered fields.

- A sentence is true of **R** if, and only if, it is derivable by first-order logic from T_{ROF}.

- There is an algorithm for deciding whether a given sentence is true of **R**, or equivalently, whether it is provable from T_{ROF}.

Details of these famous results may be found in Chang and Keisler [1973], Prestel [1984], Shoenfield [1967], and many other logic books, besides Tarski's own publications.

Ordered Affine Spaces

The first-order language L_{OS} for ordered affine geometry has a single non-logical symbol, a ternary relation symbol **B**, which is interpreted in a vector space over an ordered field as the relation B that holds of a triple (abc) when

there exists some $\lambda \in F$ with $0 \leq \lambda \leq 1$ and $b = (1 - \lambda) \cdot a + \lambda \cdot c$.

Note that this allows the relation to hold when some arguments are equal (e.g. $B(aab)$, $B(aaa)$), whereas the betweenness relation introduced in §2.7 holds only between distinct points. That relation is, however, easily defined by the L_{OS}-formula

$$B(xyz) \wedge (x \neq y) \wedge (x \neq z) \wedge (y \neq z),$$

while the present interpretation of B is simpler to work with.

Using B we now define the notions of collinearity, parallelism, coplanarity, being of dimensions 2, 3, or 4, etc.

(a). The notation $L(xyz)$ is introduced as an abbreviation for the L_{OS}-formula

$$B(xyz) \vee B(yzx) \vee B(zxy),$$

which expresses "z is collinear with x and y".

The line joining two points can then be defined as the set of points collinear with the given two. For a detailed treatment of affine planes as "collinearity structures", cf. Szmielew [1983], Chapter 2.

(b). The notation $P(x_1 x_2 x_3 z)$ abbreviates the formula $\exists y \varphi$, where φ is the disjunction of the six formulae

$$L(x_{\rho_1} x_{\rho_2} y) \wedge L(x_{\rho_3} yz),$$

where ρ is a permutation of $\{1, 2, 3\}$. This expresses "z is coplanar with x_1, x_2, and x_3", i.e. "z lies on a line which passes through a vertex of triangle $x_1 x_2 x_3$ and intersects the opposite side".

(c). Parallelism can be defined in terms of L and P: $xy \parallel zw$ abbreviates

$$P(xyzw) \wedge [\forall u(L(xyu) \rightarrow L(zwu)) \vee \forall u(L(xyu) \rightarrow \neg L(zwu))].$$

(d). The relation "z is in the threefold generated by x_1, x_2, x_3, and x_4" is expressed by the formula $TF(x_1 x_2 x_3 x_4 z)$, which is $\exists y \varphi$, where φ is the disjunction of the formulae

$$P(x_{\rho_1} x_{\rho_2} x_{\rho_3} y) \wedge L(x_{\rho_4} yz),$$

with ρ ranging over the permutations of $\{1, 2, 3, 4\}$.

(e). Similarly, using the definition of **TF**, and permutations of the set $\{1, 2, 3, 4, 5\}$, we can write down an L_{OS}-formula $\mathbf{FF}(x_1 \ldots x_5 z)$ that expresses "z is in the fourfold generated by $x_1, \ldots, x_5$".

A suitable set of axioms for the first-order theory of ordered affine fourfolds over real-closed fields is the set T_{ROS4}, comprising the universal closures of the following formulae.

1. $\mathbf{B}(xyx) \rightarrow x = y$.

2. $\mathbf{B}(xyz) \wedge \mathbf{B}(yzu) \wedge y \neq z \rightarrow \mathbf{B}(xyu)$.

3. $\mathbf{B}(xyz) \wedge \mathbf{B}(xyu) \wedge x \neq y \rightarrow \mathbf{B}(yzu) \vee \mathbf{B}(yuz)$.

4. $\exists x[\mathbf{B}(xyz) \wedge x \neq y]$.

5. $\mathbf{B}(xtu) \wedge \mathbf{B}(yuz) \rightarrow \exists v(\mathbf{B}(xvy) \wedge \mathbf{B}(ztv))$.

6. $\exists x_1 \ldots \exists x_5[\neg\mathbf{TF}(x_1 \ldots x_5) \wedge \forall z\mathbf{FF}(x_1 \ldots x_5 z)]$.

7. $\mathbf{B}(xut) \wedge \mathbf{B}(yuz) \wedge x \neq u \rightarrow \exists v\exists w[\mathbf{B}(xyv) \wedge \mathbf{B}(xzw) \wedge \mathbf{B}(vtw)]$.

8. $\exists z\forall x\forall y[\varphi \wedge \psi \rightarrow \mathbf{B}(zxy)] \rightarrow \exists u\forall x\forall y[\varphi \wedge \psi \rightarrow \mathbf{B}(xuy)]$, where φ is any formula with x free, but not y, z, or u, while ψ has y free but not x, z, or u.

These axioms are adapted from Szczerba and Tarski [1965], replacing their dimension axioms for the plane by the sentence (6) which states that the space is generated by a five-point, and so is four-dimensional, and dropping Desargues' axiom, since that becomes derivable for spaces of more than two dimensions. Axiom (5) is a version of Pasch's Law. The axiom schema (8) is a first-order version of Dedekind continuity, and it can be used in combination with "Euclid's axiom" (7) to prove the parallel postulate AS3.

The precise details of these axioms are not our main concern here: suffice it to observe that Szczerba and Tarski have shown that there is an explicitly defined set T_{ROS4} of L_{OS}-sentences for which it can be proven that

- *every model of T_{ROS4} is isomorphic to the affine ordered fourfold $\Sigma_4(F)$ over some real-closed ordered field F.*

The above analysis of the theory T_{ROF} can then be used to show that T_{ROS4} is deductively complete and decidable. Each sentence σ of L_{OS} is translated into a sentence σ^+ of L_{OF}, by formalising the definitions of equality and betweenness of points in $\Sigma_4(F)$, these definitions themselves being given in terms of the field structure of

F. First, each variable x is assigned a list $(x_1, \ldots, x_4)$ of four distinct variables. Then, for atomic formulae, $(x = y)^+$ is defined to be

$$(x_1 = y_1) \wedge \ldots \wedge (x_4 = y_4),$$

while $\mathbf{B}(xyz)^+$ is

$$\exists \lambda [(0 \leq \lambda) \wedge (\lambda \leq 1) \wedge \varphi_1 \wedge \ldots \wedge \varphi_4],$$

where φ_i is

$$y_i = (1 - \lambda) \cdot x_i + \lambda \cdot z_i.$$

Each universal quantifier $\forall x$ in σ is replaced by $\forall x_1 \forall x_2 \forall x_3 \forall x_4$, an existential quantifier $\exists x$ is replaced by $\exists x_1 \exists x_2 \exists x_3 \exists x_4$, and the propositional connectives $\wedge$, $\vee$, $\neg$, $\rightarrow$, $\leftrightarrow$, are left as they are. It can then be seen that for any ordered field F,

- σ is true of $\Sigma_4(F)$ iff σ^+ is true of F.

Therefore, if F is real-closed, so that the sentences true of F are precisely those that are true of $\mathbf{R}$, it follows that

- σ is true of $\Sigma_4(F)$ iff σ^+ is true of $\mathbf{R}$,

and indeed

- σ is true of $\Sigma_4(F)$ iff σ is true of $\Sigma_4(\mathbf{R})$.

Thus all models of T_{ROS4} satisfy the same L_{OS}-sentences, so such a sentence σ is either true of all T_{OS}-models, or false in all such models. By the Completeness Theorem of first-order logic, it follows that σ is either provable or refutable (its negation is provable) from T_{ROS4}, and especially, that

- σ is true of $\Sigma_4(\mathbf{R})$ iff it is derivable by the rules of first-order logic from T_{ROS4}.

Also, since truth of σ^+ in $\mathbf{R}$ is algorithmically decidable, and an explicit procedure was given for translating σ into σ^+, it follows that there is an algorithm for deciding whether or not a given L_{OS}-formula σ is true of $\Sigma_4(\mathbf{R})$ (or equivalently, provable from T_{ROS4}).

It is noteworthy in passing that the quantifier-elimination method used by Tarski for the theory of real-closed fields does not apply to T_{ROS4} itself. To see this, let $\varphi(xyzu)$ be the L_{OS}-formula

$$\exists w [(xy \parallel zw) \wedge (xz \parallel yw) \wedge (zy \parallel wu)].$$

Suppose that x, y, and z denote three non-collinear points in $\Sigma_4(\mathbb{R})$, and that in the plane they generate, these points play the role of the points o, e, and e', respectively, in defining an addition operation $+$ on the line xy, by the method described in Figure 2.8 of §2.1. Then if u denotes the point $y + y$ (Figure A),

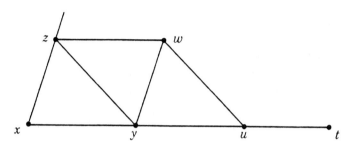

Figure A

we have that

$$\varphi(xyzu) \text{ is satisfied.} \tag{i}$$

But if t denotes the point $y + u$ (i.e. $y + y + y$), then

$$\varphi(xyzt) \text{ is not satisfied,} \tag{ii}$$

where $\varphi(xyzt)$ is the result of replacing u by t in $\varphi(xyzu)$.

Now if T_{ROS4} did admit elimination of quantifiers, there would be a quantifier-free formula $\psi(xyzu)$, having only the variables x, y, z, and u, such that the sentence

$$\forall x \forall y \forall z \forall u [\varphi(xyzu) \leftrightarrow \psi(xyzu)] \tag{iii}$$

was provable from T_{ROS4}, hence true of $\Sigma_4(\mathbb{R})$. But then for the interpretation of our example, (i)-(iii) yield

$$\psi(xyzu) \text{ is satisfied,} \tag{iv}$$

and

$$\psi(xyzt) \text{ is not satisfied.} \tag{v}$$

But this is impossible, since there is an evident isomorphism f between the two substructures

$$(\{x, y, z, u\}, B) \quad \text{and} \quad (\{x, y, z, t\}, B)$$

of $\Sigma_4(\mathbb{R})$, namely the map that fixes x, y, and z, and has $f(u) = t$ (in each structure, B is the linear betweenness relation). From this it follows that for any quantifier-free formula $\psi(xyzu)$ of L_{OS},

$\psi(xyzu)$ is satisfied in $\Sigma_4(\mathbf{R})$ iff $\psi(f(x)f(y)f(z)f(u))$ is satisfied in $\Sigma_4(\mathbf{R})$,

which is contrary to (iv) and (v).

(The distinction between variables and the elements they denote has been deliberately obscured in this example, to simplify the presentation, but the situation should be clear enough.)

Metric Ordered Spaces

The language L_{MOS} for metric ordered affine spaces is obtained by adding to L_{OS} a four-placed relation symbol $\perp$. Associated atomic formulae are written $(xy \perp zw)$ and are intended to express the relation "the line joining x to y is orthogonal to the line joining z to w", i.e. "$(x - y) \bullet (z - w) = 0$". Translation of this into an L_{OF}-formula will depend on which inner product is being used to define orthogonality.

First-order axioms for metric ordered affine fourfolds are obtained by adding to the theory T_{ROS4} sentences expressing the axioms OS1 - OS5 of §4.2. For these the notation $x\lambda y$ will be used to stand for the formula

$$x \neq y \land xy \perp xy,$$

expressing "xy is a null line". The axioms are as follows.

OS1. $xy \perp zw \rightarrow zw \perp xy$.

OS2. $[\mathbf{P}(xyzw) \rightarrow xy \perp zw] \lor$
$\qquad \exists t[\mathbf{P}(xyzt) \land \forall u(\mathbf{P}(xyzu) \rightarrow (xy \perp zu \leftrightarrow \mathbf{L}(tzu)))]$,

   i.e. in the plane of x, y, and z, either xy is singular, or else there is exactly one line through z orthogonal to xy.

OS3. $[xy \perp zw \land xz \perp yw] \rightarrow xw \perp yz$.

OS4. $[xy \perp yw \land xy \perp yz \land \mathbf{P}(ywzu)] \rightarrow xy \perp yu$.

OS5. $[xy \perp zw \land zw \parallel uv] \rightarrow xy \perp uv$.

An axiom set T_{M4} for Minkowski spacetime is obtained by adding to T_{ROS4} the universal closures of these formulae OS1 - OS5, as well as

M1. $\forall x \forall y \exists w \neg(xy \perp yw)$
   (no singular lines),

M2. $\exists x \exists y (x \lambda y)$
   (there exist null lines),

M3. $[x\lambda y \wedge z\lambda w \wedge xy \perp zw] \rightarrow xy \parallel zw$
   (no intersecting orthogonal null lines).

Now any model of T_{M4} is, in particular, a model of T_{ROS4}, and so is coordinatisable as the affine space $\Sigma_4(F)$ over some real-closed ordered field F. Since the model satisfies OS1 - OS5 and M1, the work of §5.3 produces an inner product on F^4 characterising the orthogonality relation interpreting $\perp$. But F is quadratic, and so the analysis of §5.4 reduces this inner product to one of three possibilities. Since the axioms M2 and M3 hold, this must in fact be the Minkowskian inner product. Thus

- *every model of T_{M4} is isomorphic to the Minkowskian spacetime $M_4(F)$ over some real-closed field F.*

The definition of the Minkowskian inner product allows the translation of L_{OS}-formulae into L_{OF}-formulae to be lifted to L_{MOS}, by defining $(xy \perp zw)^+$ to be the formula

$$u_1 \cdot v_1 + u_2 \cdot v_2 + u_3 \cdot v_3 - u_4 \cdot v_4 = 0,$$

where u_i is $(x_i - y_i)$ and v_i is $(z_i - w_i)$. Then for any real-closed field F,

- *an L_{MOS}-sentence σ is true of $M_4(F)$ iff σ^+ is true of* **R**.

From this it follows that

- *σ is true of real Minkowskian spacetime $M_4(\mathbf{R})$ iff it is derivable by first-order logic from T_{M4},*

and

- *the question "is σ true of Minkowskian spacetime?" is algorithmically decidable.*

Axioms for Euclidean four-space arise by replacing M1 - M3 in T_{M4} by

$$\forall x \forall y \neg (x\lambda y),$$

and the proof that the resulting theory is deductively complete and decidable follows by using the Euclidean inner product as the definition of the translation $(xy \perp zw)^+$.

A similar analysis can be carried out for Artinian four-space, and indeed for all the other metric affine geometries that we have characterised in this study.

Appendix B

After and the Alexandrov-Zeeman Theorem

The first qualitative axiomatisation of Minkowskian spacetime was set out by A.A.Robb in his book *A Theory of Time and Space* [1914] (revised as *Geometry of Time and Space* [1936]), which gave a categorical set of axioms having the binary relation "after" as its only primitive notion. In this relation, y is *after* x if y lies inside or on the future light cone based at x. This means that is possible to send a signal from location x to location y at a velocity less than or equal to that of light.

Robb's work was seen to be even more remarkable upon the subsequent discovery (Tarski [1935], Robinson [1959]) that no collection of binary relations can provide an adequate primitive basis for Euclidean geometry. However his approach has never gained much popularity, perhaps because many of the twenty-one postulates he used seem ad hoc, introduced only as needed to allow the technical development of the theory to proceed. These postulates generally lack physical significance, or even a more general geometrical intuitiveness, and as Suppes [1973] observes, "the complexity of the axioms stands in marked contrast to the simplicity of his single primitive concept".

Nonetheless, the fact that such a natural primitive notion does suffice is an important contribution to an understanding of the geometry of spacetime. In this Appendix it will be shown how to recover Robb's result, by showing that the primitives *between* and *orthogo-*

nal that we have used can be defined in terms of *after*. As a bonus, this analysis will provide easy access to the famous Alexandrov-Zeeman characterisation of spacetime transformations.

In Minkowskian spacetime, a point $y = (y_1, \ldots, y_4)$ is *after* point $x = (x_1, \ldots, x_4)$, denoted $y(after)x$, when

$$(y - x)^2 \leq 0 \quad \text{and} \quad x_4 < y_4$$

(where in general z^2 means the Minkowskian inner product $z \bullet z$).

We now build up a sequence of definitions of relations between points, using the standard logical symbols $\neg$ (not), $\wedge$ (and), $\vee$ (or), $\rightarrow$ (implies), $\forall$ (for all), $\exists$ (there exists).

After at Light-Speed

$x \ll y$ means

$$\exists z[z \neq x \wedge \neg(z(after)x \vee x(after)z) \wedge y(after)x \wedge y(after)z \wedge$$
$$\forall u(\, u(after)x \wedge u(after)z \rightarrow \neg y(after)u\,)].$$

This states that there is a point z, neither before nor after x, such that y is after both x and z but not after any other point that is after x and z. This means that y lies on the future light cone of x, i.e. the line xy is lightlike (null), with y after x. If α is an inertia (isotropic) plane containing xy, then z can be taken as any point in the past of y that lies on the other lightlike line through y in α. The definition of $x \ll y$ expresses the fact that in α, y is the *least upper bound* of x and y under the ordering relation *after*.

Lightlike Connectibility

$x \lambda y$ means

$$(x \ll y) \vee (y \ll x),$$

which is satisfied precisely when the line joining x to y is lightlike.

Notice that the relation λ is more "basic" than *after*, since, whereas λ is symmetric, *after* gives the "direction of time", and so cannot be defined in terms of λ.

All of the definitions to follow will actually be built out of the relation λ, showing that λ is itself a suitable primitive for spacetime geometry (Latzer [1973]).

Lightlike Collinearity
$\lambda(xyz)$ means

$$x\lambda y \wedge y\lambda z \wedge x\lambda z.$$

This holds if, and only if, x, y, and z all lie on the same lightlike line. For, if there were a triangle xyz of lightlike lines, then in the plane they generate xy, xz, and the line through x parallel to yz would form three distinct concurrent null lines. But this can only happen in a null plane, and spacetime has none of those.

Optical Plane Generation
(orthogonality for spacelike and lightlike lines). $S(xyz)$ means

$$x\lambda y \wedge \neg\lambda(xyz) \wedge \forall u(z\lambda u \rightarrow \neg\lambda(xyu)).$$

This holds when x, y, and z are not collinear, and in the plane they generate xy is a null line that is not intersected by any null line through z. Thus $S(xyz)$ iff x, y, and z generate an optical plane in which line xy is lightlike.

It follows that $S(xyz)$ iff xy is lightlike, xz is spacelike, and the two are orthogonal. So we already have a definition of one case of the relation of orthogonality.

Spacelike Connectibility
$x\sigma z$ means

$$\exists y S(xyz).$$

As noted above, $S(xyz)$ implies xz is spacelike. Since every spacelike line in spacetime lies in some optical plane, $x\sigma z$ holds precisely when xz is a spacelike line.

Spacelike Collinearity
$\sigma(xyz)$ means

$$x\sigma y \wedge y\sigma z \wedge x\sigma z \wedge \neg\exists u(x\lambda u \wedge y\lambda u \wedge z\lambda u).$$

This states that the lines xy, yz, and xz are all spacelike, and that the lightcones at x, y, and z do not have a common point of intersection. In spacetime, this holds precisely when the points x, y, and z lie on the same spacelike line.

It is straightforward to see that if x, y, and z are collinear and spacelike then there can be no point u on all their lightcones. For otherwise, ux, uy, and uz would be three null lines through u in the plane containing these points. But, as has already been noted, no plane in spacetime has such a configuration. The proof that, conversely, $\sigma(xyz)$ implies that x, y, and z are collinear is due to Latzer [1973], and uses the following intuitively appealing physical argument.

Suppose that $\sigma(xyz)$ holds, but x, y, and z are not collinear. Then since $x\sigma y$, we can choose a reference frame (coordinatisation) in which x and y are simultaneous. Let X, Y, Z be the spatial positions, and t_x, t_y, t_z the time coordinates, of x, y, z, respectively, in this reference frame. Let M be the perpendicular bisector of XY in the plane of X, Y, Z (Figure B.1).

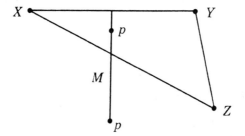

Figure B.1

Take any position p on M and let light signals be sent toward p from X and Y at time $t_x = t_y$, and from Z at time t_z. The signals from X and Y arrive at p at the same time t_p. If there are positions p at which the signal from Z arrives earlier than t_p, and other positions at which this signal arrives later than t_p, then by the Intermediate Value Theorem there will be a position p on M at which all three signals arrive simultaneously. Then the spacetime location $u = (p, t_p)$ lies on all three lightcones, contradicting $\sigma(xyz)$.

The rest of the proof is a consideration of cases, depending on the relative positions of X, Y, Z, and M, to show that there do exist positions p as required.

It is not technically essential to use the Intermediate Value Theorem here, and one could give a proof that is more algebraic (and perhaps less intuitive) by solving the equations of the lightcones to obtain an intersection point. These equations are quadratic, and so require only the extraction of square roots (in addition to field operations) for their solution. In fact the definition of $\sigma(xyz)$ characterises spacelike collinearity in the Minkowskian spacetime over any quadratic ordered field, as has been verified in Kelley [1985].

Spacelike Plane Generation
$P_\sigma(xyzw)$ means

$$x\sigma y \wedge y\sigma z \wedge x\sigma z \wedge \neg\sigma(xyz) \wedge \exists u \exists v[\sigma(uvw)\wedge$$
$$[\{\sigma(xyu) \wedge (\sigma(xzv) \vee \sigma(yzv))\} \vee \{\sigma(xzu) \wedge \sigma(yzv)\}]].$$

This states firstly that xyz is a triangle of spacelike lines, and secondly that there is a spacelike line through w meeting two sides of this triangle. This implies that w lies in the plane of the triangle. But any point in this plane lies on a spacelike line meeting two sides of the triangle: just take a line parallel to one of the sides.

Thus $P_\sigma(xyzw)$ holds iff x, y, and z form a spacelike triangle and w is a point in the plane they generate.

Coplanarity
$P(xyzw)$ means

$$\exists t \exists u \exists v[P_\sigma(tuvx) \wedge P_\sigma(tuvy) \wedge P_\sigma(tuvz) \wedge P_\sigma(tuvw)],$$

stating that there is a spacelike triangle whose plane contains the points x, y, z, and w. Since in Minkowskian spacetime every plane contains spacelike triangles, $P(xyzw)$ holds iff x, y, z, and w are coplanar.

Collinearity
$L(xyz)$ means

$$\exists u \exists v[P(xyzu) \wedge P(xyzv) \wedge \neg P(xyuv)],$$

stating that x, y, and z lie in the intersection of two distinct planes, i.e. are collinear.

Parallelism
$xy \parallel zw$ means

$$P(xyzw) \wedge [\forall u(L(xyu) \to L(zwu)) \vee \forall u(L(xyu) \to \neg L(zwu))].$$

This is the same definition of parallelism in terms of coplanarity and collinearity that was mentioned in Appendix A.

Timelike Connectibility
$x\tau y$ means
$$(x \neq y) \wedge \neg(x\lambda y) \wedge \neg(x\sigma y),$$
and holds iff xy is a timelike line.

Timelike Collinearity

$\tau(xyz)$ means

$$L(xyz) \wedge \neg\lambda(xyz) \wedge \neg\sigma(xyz),$$

and holds iff x, y, and z all lie on a timelike line.

Linear Betweenness

To define $B(xyz)$, we first define the relation $B_\tau(xyz)$, of *betweenness on a timelike line*, to mean

$$\tau(xyz) \wedge \forall w(y\tau w \rightarrow x\tau w \vee z\tau w).$$

This asserts that x, y, and z lie on a timelike line, with the "timecone" of y being contained within the union of the timecones of x and z. In the two-dimensional illustration of Figure B.2, the forbidden region for any w such that $y\tau w$ is shaded. Any point on xz not between x and z is connected by timelike lines to points in this shaded region.

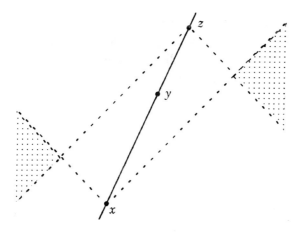

Figure B.2

Now we define $B(xyz)$ to mean

$$L(xyz) \wedge \exists u \exists v[\, B_\tau(xuv) \wedge uy \parallel vz\,].$$

This specifies betweenness on a general line xyz by parallel projection from an intersecting timelike line (Figure B.3). In any ordered affine plane, parallel projection preserves linear betweenness (this statement is actually a version of Pasch's Law: cf. Szmielew [1983], p.148).

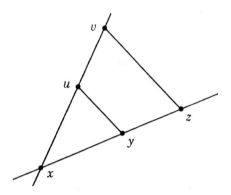

Figure B.3

The definition of B_τ itself is a special case of the condition

$$x\tau y \wedge y\tau z \wedge x\tau z \wedge \forall w(y\tau w \to x\tau w \vee z\tau w),$$

which was shown by Latzer [1973] to characterise the general (non-linear) temporal betweenness relation that holds when it is possible to pass through y while travelling between x and y at slower-than-light velocity.

Optical Parallelogram
$O(xyzw)$ means

$$x\lambda z \wedge x\lambda w \wedge \neg\lambda(xzw) \wedge xz \parallel yw \wedge xw \parallel yz,$$

and states that xy and zw are the diagonals of a parallelogram whose sides are all lightlike. The plane containing this parallelogram must be inertial (isotropic), as it has intersecting null lines (xz and xw). In Theorem 2.5.1 it was shown that the diagonals of a parallelogram of this type are orthogonal.

Orthogonality in an Inertia Plane
$I(xyzw)$ means

$$\exists u \exists v[O(xyuv) \wedge uv \parallel zw \wedge P(xyzw)],$$

which implies that there is an optical parallelogram whose plane contains zw, and whose diagonals are xy and a line uv parallel to zw. In view of the discussion of the relation O, it follows that the plane is inertial, that lines xy and zw are orthogonal, and hence that one is timelike and the other spacelike.

Now in any inertia plane, if line xy is either spacelike or timelike, then by intersecting the lightlike lines through x and y, an optical parallelogram is obtained with xy as one diagonal, and with the other diagonal being orthogonal to xy, and so being parallel to any line orthogonal to xy in this plane. Thus the relation $I(xyzw)$ holds precisely when there exists an inertia plane in which xy and zw are non-null orthogonal lines.

Orthogonality
$xy \perp zw$ means

$$\exists u[xu \parallel zw \wedge (\varphi_1 \vee \varphi_2 \vee \varphi_3)],$$

where

φ_1 is $x\lambda y \wedge [\lambda(xyu) \vee S(xyu)]$,

φ_2 is $x\tau y \wedge I(xyxu)$,   and

φ_3 is $\exists t \exists v[S(xty) \wedge S(xvy) \wedge \neg\lambda(xtv) \wedge P(xtvu)]$.

This definition uses parallelism to reduce the characterisation of lines orthogonal to xy to those lines xu passing through x. The disjuncts φ_1, φ_2, and φ_3 cover all possibilities as to the nature of xy.

φ_1: If xy is null, then xu is orthogonal to xy iff either they are the same null line ($\lambda(xyu)$), or else xu is spacelike and the plane containing x, y, and u is optical ($S(xyu)$).

φ_2: If xy is timelike, then any plane containing xy will be inertial, and our discussion of the relation I implies that xy will be orthogonal to xu iff $I(xyxu)$.

φ_3: If xy is spacelike, then the threefold Σ through x orthogonal to xy will be an inertia threefold (a copy of three-dimensional spacetime). If xy is orthogonal to xu, then xu lies in Σ and so lies in some inertia plane α in Σ. Then if xt and xv are the two null lines through x in α (Figure B.4), xt and xv will each be orthogonal to xy, so $S(xty)$ and $S(xvy)$, and hence φ_3 is satisfied.

Conversely, if t and v fulfill φ_3, then xy is spacelike, and orthogonal to the distinct null lines xt and xv, so is orthogonal to all lines in the plane of x, t, and v, and in particular is orthogonal to xu.

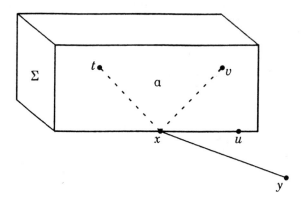

Figure B.4

This completes the demonstration that the relations of linear betweenness and orthogonality are definable in terms of lightlike connectibility λ, and hence in terms of the relation *after*. The demonstration has been largely descriptive, and the reader who would like
a more rigorous presentation may be interested in verifying that the
definitions work in the Minkowskian spacetime over any quadratic
ordered field F. In fact they work in the n-dimensional version of
spacetime, for $n \geq 3$, given by the inner product

$$x_1 \cdot y_1 + \cdots + x_{n-1} \cdot y_{n-1} - x_n \cdot y_n$$

on F^n. In the case $n = 2$, the definitions do not work: in particular
the definitions of $S(xyz)$ and $x\sigma z$ do not make sense. Indeed, a
rotation of the Lorentz plane through 45° will map the light-cone
back onto itself, while interchanging timelike and spacelike vectors,
indicating that σ and τ are not definable in terms of λ in that plane.

The Alexandrov-Zeeman Theorem

The "causal" ordering $x \prec y$ is defined from the relation *after* by
the condition

$$y(after)x \wedge \neg x \ll y,$$

which means that y lies *inside* the future light-cone of x, i.e. that
a slower-than-light signal can be sent from x to y. Conversely,
$y(after)x$ is definable from $\prec$ by the condition

$$x \neq y \wedge \forall z[y \prec z \rightarrow x \prec z].$$

A *causal automorphism* of spacetime is a bijection $f : \mathbb{R}^4 \to \mathbb{R}^4$ satisfying

$$x \prec y \quad \text{iff} \quad f(x) \prec f(y).$$

In an influential paper "Causality implies the Lorentz group", E.C. Zeeman [1964] proved that any causal automorphism of spacetime is a *Lorentz transformation*, up to a dilation and a translation. A Lorentz transformation is an automorphism of spacetime as a metric vector space, i.e. a linear bijection $g : \mathbb{R}^4 \to \mathbb{R}^4$ that preserves the Minkowskian inner product, in the sense that

$$g(x) \bullet g(y) = x \bullet y.$$

These transformations are the cornerstone of special relativity theory, since they are precisely the mappings that translate the coordinate system (reference frame) of one observer into that of another.

Zeeman's result states that if f is a causal automorphism, then there is a Lorentz transformation g, a positive constant c, and a vector v, such that

$$f(x) = cg(x) + v, \tag{1}$$

so that f is the composition of g followed by the *dilation* $y \mapsto cy$, and then the *translation* $z \mapsto z + v$. Since g is linear, in fact $v = f(\mathbf{o})$.

This result generated a good deal of research into spacetime transformations. It transpired that the phenomenon had already been discovered in Russia in 1949 by A.D.Alexandrov (cf. Alexandrov [1949, 1967], Alexandrov and Ovchinnikova [1953]), and similar results were obtained in China in the 1950's by Loo-Keng Hua (cf. Hua [1981]). Alexandrov derived the representation (1) for any bijection that maps light-cones onto light-cones. Such a mapping may alternatively be defined as a *λ-automorphism*, i.e. a bijection $f : \mathbb{R}^4 \to \mathbb{R}^4$ satisfying

$$x\lambda y \quad \text{iff} \quad f(x)\lambda f(y), \tag{2}$$

where λ is the lightlike connectibility relation defined above in terms of *after*. Since *after* is definable from $\prec$, every causal automorphism is a λ-automorphism, so Zeeman's result follows from Alexandrov's (with the additional conclusion in the causal case that g is orthochronous (time preserving)).

All approaches to deriving (1) proceed by showing that f maps lines onto lines, for the rest of the argument reduces to standard theory, which we now outline.

Lemma 1. *If* $f : \mathbb{R}^4 \to \mathbb{R}^4$ *is a bijection that maps lines onto lines, then* f *is linear up to a translation, i.e. there is a linear bijection* $h : \mathbb{R}^4 \to \mathbb{R}^4$ *such that*

$$f(x) = h(x) + f(\mathbf{o}). \tag{3}$$

Proof (outline). This is a special case of the so called Fundamental Theorem of Affine Geometry (cf., e.g., Snapper and Troyer [1971], §§15-18), which states that the theorem holds for any vector space of dimension two or more, but with the weaker conclusion that h is "semilinear with respect to an automorphism of the scalar field". However the field $\mathbb{R}$ has no automorphisms other than the identity function, and so the form (3) obtains. The idea of the proof is natural: define

$$h(x) = f(x) - f(\mathbf{o}),$$

and show that h is linear. To prove that h preserves vector sums and scalar multiples, it is necessary to show that it maps planes onto planes, and parallelograms onto parallelograms etc., and these follow from the fact that f preserves lines. $\square$

Lemma 2. *If* h *is a linear* λ-*automorphism, there exists a positive real number* d *such that*

$$h(x) \bullet h(y) = d(x \bullet y). \tag{4}$$

Proof (sketch). Let $e = (0,0,0,1)$, and $d = -(h(e)^2)$ (where $z^2 = z \bullet z$ as usual). Then if u is any vector in the Euclidean sphere

$$S = \{u \in \mathbb{R}^4 : u_4 = 0 \text{ and } u \bullet u = 1\},$$

algebraic calculations show that

$$h(e) \bullet h(u) = 0 \quad \text{and} \quad h(u)^2 = d. \tag{5}$$

This uses linearity, and the fact that $e + u$ and $e - u$ are null, so their h-images are null (since $h(\mathbf{o}) = \mathbf{o}$, $\mathbf{o}\lambda z$ implies $\mathbf{o}\lambda h(z)$ etc.).

Next, any vector z can be written in the form $\lambda e + \mu u$ with $u \in S$, from which by (5), using $e^2 = -1$ and $u^2 = 1$, it may be shown that $h(z)^2 = d(z^2)$. Since in general

$$x \bullet y = \frac{1}{2}((x + y)^2 - x^2 - y^2),$$

further calculations then give (4).

To show that d is positive, take any separation (spacelike) plane through the origin and observe that the h-image of this plane is also a plane through the origin, and so itself contains spacelike vectors (as all planes have these). Hence there exist spacelike vectors z with $h(z)$ also spacelike, so that z^2 and $h(z)^2$ are both positive. Since $h(z)^2 = d(z^2)$, the result follows. □

It is noteworthy, in passing, that there is an alternative *topological* argument showing that the constant d in this Lemma must be positive. For, since in general $h(z)^2 = dz^2$, if d were negative then h would map spacelike vectors to timelike ones, and vice versa. Thus the *spacecone* $\{z : z^2 > 0\}$ would be mapped by h onto the *timecone* $\{z : z^2 < 0\}$. But in spacetime of three or more dimensions, the spacecone is *connected* in the usual (Euclidean!) topology (cf. Figure 1.8), while the timecone is disconnected, having the future and the past as disjoint components. But h is continuous, being linear, and so cannot map a connected set onto a disconnected one.

This point underlines the difference between the Lorentz plane and spacetimes of more than one spatial dimension, and helps to explain why the Alexandrov-Zeeman theorem fails in the Lorentz plane, where there is a symmetry between the cones of space and time.

Corollary 3. *If f is a λ-automorphism mapping lines onto lines, then f can be represented in the form (2), i.e.*

$$f(x) = cg(x) + f(\mathbf{o}),$$

where g is a Lorentz transformation, and c a positive constant.

Proof. By Lemma 1, there is a linear bijection h with

$$f(x) = h(x) + f(\mathbf{o}).$$

Then $h(x) - h(y) = f(x) - f(y)$, so h is a λ-automorphism since f is. Hence by Lemma 2,

$$h(x) \bullet h(y) = d(x \bullet y) \tag{4}$$

for some positive constant d.

Now let $c = \sqrt{d}$, and define

$$g(x) = \frac{1}{c}h(x).$$

Then g is a linear bijection which, from (4), satisfies

$$g(x) \bullet g(y) = x \bullet y,$$

so g is a Lorentz transformation. Since $cg(x) = h(x)$, the proof is complete. $\square$

Thus the essence of the Alexandrov-Zeeman theorem consists in showing that any λ-automorphism must map lines onto lines. Proofs of this have been given that are geometric, topological, or algebraic, but the first part of this Appendix suggests a proof that might be called *logical*, or *definitional*. We saw how the collinearity relation L can be defined in terms of λ. The definition of $L(xyz)$, when written out in full, is a (rather long) statement involving only λ and logical symbols (symbols of first-order logic, to be precise). It follows that any bijection f satisfying

$$x\lambda y \quad \text{iff} \quad f(x)\lambda f(y) \tag{2}$$

will satisfy

$$L(xyz) \quad \text{iff} \quad L(f(x)f(y)f(z)) \tag{6}$$

(this is an instance of the fact that the equivalence

$$\varphi(x_1,\ldots,x_n) \quad \text{iff} \quad \varphi(f(x_1),\ldots,f(x_n))$$

is satisfied for any statement φ in the first-order language of λ).
    But any bijection satisfying (6) must map lines onto lines. $\square$

Bibliography

Alexandrov, A.D.
[1949] On Lorentz transformations, *Sessions of the Mathematical Seminar of the Leningrad Section of the Mathematical Institute*, 15 September 1949 (abstract, in Russian).

[1967] A contribution to chronogeometry, *Canadian J. Math.* **19**, 1119-1128.

Alexandrov, A.D., and Ovchinnikova, V.V.
[1953] Notes on the foundations of relativity theory, *Vestnik Leningrad University* **11**, 95-100 (in Russian).

Artin, E.
[1957] *Geometric Algebra*, Wiley.

Baer, Reinhold
[1944] The fundamental theorems of elementary geometry: an axiomatic analysis, *Trans. Amer. Math. Soc.* **56**, 94-129.

Behnke, H., Bachmann, F., Fladt, K., and Kunle, H. (eds.)
[1974] *Fundamentals of Mathematics, vol. II: Geometry*, MIT Press.

Bennett, M.K.
[1973] Coordinatisation of affine and projective spaces, *Discrete Math.* **4**, 219-231.

Birkhoff, Garrett, and Maclane, Saunders
[1965] *A Survey of Modern Algebra, Third Edition*, MacMillan.

Blumenthal, L.M.
[1961] *A Modern View of Geometry*, W.H. Freeman.

Chang, C.C., and Keisler, H.J.
[1973] *Model Theory*, North-Holland.

Coxeter, H.S.M.
[1942] *Non-Euclidean Geometry*, University of Toronto Press.

[1949] *The Real Projective Plane*, McGraw-Hill.

[1964] *Projective Geometry*, Blaisdell.

Ewald, Gunter
[1971] *Geometry: An Introduction*, Wadsworth.

Garner, Lynne E.
[1981] *An Outline of Projective Geometry*, North-Holland.

Hartshorne, Robin
[1967] *Foundations of Projective Geometry*, W.A. Benjamin.

Henkin, Leon; Suppes, Patrick; and Tarski, Alfred (eds.)
[1959] *The Axiomatic Method*, North-Holland.

Hilbert, David
[1971] *Foundations of Geometry, Tenth Edition*, The Open Court.

Hua, Loo-Keng
[1981] *Starting with the Unit Circle*, Springer-Verlag.

Kelly, Susan M.
[1985] *Characterisations of Lorentz Transformations*, M.A. thesis, Victoria University of Wellington.

Latzer, Robert W.
[1973] Non-directed light signals and the structure of time, in *Suppes* [1973], 321-365.

Mihalek, R.J.
[1972] *Projective Geometry and Algebraic Structures*, Academic Press.

Prestel, Alexander
[1984] *Lectures on Formally Real Fields*, Lecture Notes in Mathematics **1093**, Springer-Verlag.

Robb, A.A.
[1914] *A Theory of Time and Space*, Cambridge University Press. Revised edition, *Geometry of Time and Space*, published in 1936.

Robinson, Raphael M.
[1959] Binary relations as primitive notions in elementary geometry, in *Henkin et. al.* [1959], 68-85.

Sasaki, Usa
[1952] Lattice theoretic characterisation of affine geometry of arbitrary dimensions, *Hiroshima Univ. J. of Science, Series A* **16**, 223-238.

Schur, F.
[1903] Zur proportionslehre, *Mathematische Annalen* **57**, 205-208.

Seidenberg, A.
[1962] *Lectures in Projective Geometry*, van Nostrand.

Shoenfield, Joseph R.
[1967] *Mathematical Logic*, Addison-Wesley.

Snapper, Ernst, and Troyer, Robert J.
[1971] *Metric Affine Geometry*, Academic Press.

Stevenson, Frederick W.
[1972] *Projective Planes*, W.H. Freeman.

Struik, Dirk J.
[1953] *Lectures on Analytic and Projective Geometry*, Addison-Wesley.

Suppes, Patrick (ed.)
[1973] *Space, Time, and Geometry*, D.Reidel.

Szczerba, L., and Tarski, A.
[1965] Metamathematical properties of some affine geometries, in *Proc. 1964 International Congress for Logic, Methodology, and Philosophy of Science*, ed. by Y. Bar-Hillel, North-Holland, 166-178.

[1979] Metamathematical discussion of some affine geometries, *Fundamenta Mathematicae* CIV, 155-192.

Szmielew, Wanda
[1983] *From Affine to Euclidean Geometry*, D.Reidel.

Tarski, Alfred
[1951] *A Decision Method for Elementary Algebra and Geometry*, second edition (revised), Univ. of California Press.

[1959] What is elementary geometry?, in *Henkin et. al.* [1959], 16-29.

[1967] *The Completeness of Elementary Algebra and Geometry*, Institute Blaise Pascal, Paris.

Tarski, A., and Lindenbaum A.
[1935] Über die Beschränktheit der Ausdrucksmittel deduktiver Theorien, *Ergebnisse eines Mathematischen Kolloquiums*, Univ. Vienna **7**, (1934/5), 15-22. English translation as "On the limitations of the means of expression of deductive theories", in *A.Tarski, Logic, Semantics, Metamathematics, Papers from 1923-1938*, Oxford 1956.

Veblen, Oswald, and Young, John Wesley
[1910] *Projective Geometry, vol. I*, Ginn.

Zeeman, E.C.
[1964] Causality implies the Lorentz group, *Journal of Mathematical Physics* **5**, 490-493.

Index

This book was typeset by the author in Computer Modern eleven-point type, using the program TeX created by Donald Knuth of Stanford University, who also designed the Computer Modern fonts. The book design is by Dikran Karagueuzian and the author.

TeX was run on a DEC 20 computer, and the output printed on an Imagen 8/300 laser printer, with a resolution of 300 dots per inch. The diagrams were generated on a Xerox 1108 (Dandelion) computer, using the Interlisp drawing system Sketch, which was designed and developed by Richard Burton at the Xerox Palo Alto Research Center.

Production of camera-ready copy took place at Stanford University's Center for the Study of Language and Information, using facilities made available in part through an award to the Center from the System Development Foundation.